THE HUMAN GENOME MAPPING PROJECT IN THE UK
PRIORITIES AND OPPORTUNITIES IN GENOME RESEARCH

A report commissioned by the Office of Science and Technology to identify areas in human genome research important for investment in the UK to benefit both basic science and commercial opportunities.

April 1994

LONDON: HMSO

PREFACE

In response to the growing importance of this burgeoning area of science, I decided, as Chief Scientific Adviser to the Government, to establish an Advisory Committee on Human Genome Research. This Committee, which I chair, was set up in January 1993. It comprises representatives of Government, the Research Councils, academia, the research charities and industry. The Committee is charged with providing advice to the Chancellor of the Duchy of Lancaster and the Office of Science and Technology on how best to build on the existing strengths of UK human genome research and secure the benefits it promises for human welfare, health care and UK industry.

It was agreed that it was important to obtain the views of scientists working at the cutting edge of human genome research in the UK. The Committee therefore convened an Expert Working Group of leading geneticists, chaired by Professor Kay Davies, to review this area and identify priorities for investment in the UK to benefit both basic science and commercial opportunities. This report presents the findings of this independent Group. It is addressed to all funding bodies with interests in UK human genome research: those supported by Government; the research charities and the private sector.

I am most grateful to all who have contributed their time and effort to this important review.

WILLIAM D P STEWART

Chief Scientific Adviser

CONTENTS

MEMBERS OF THE EXPERT WORKING GROUP

Professor Kay E Davies (Chairman)
Director, MRC Clinical Sciences Centre,
Royal Postgraduate Medical School, London

Professor John Bell,
Nuffield Professor of Clinical Medicine, John Radcliffe Hospital, Oxford

Professor Martin Bobrow FRCP FRCPath,
Prince Philip Professor of Paediatric Research, Guy's Hospital, London

Professor Graham Boulnois,
Chief Biotechnologist, Zeneca Pharmaceuticals plc

Dr Richard Durbin,
Joint Head of Informatics, Sanger Centre, and
MRC Laboratory of Molecular Biology, Cambridge

Dr Anna-Maria Frischauf,
Head, Molecular Analysis of Mammalian Mutation Laboratory,
Imperial Cancer Research Fund, London

Professor Peter Goodfellow FRS,
Balfour Professor of Genetics, University of Cambridge

Dr Tim Harris,
Director of Biotechnology, Glaxo Group Research plc

Professor Nick Hastie,
Assistant Director, MRC Human Genetics Unit, Edinburgh

Dr Peter Little,
Department of Biochemistry, Imperial College London

Dr John Sulston FRS,
Director, Sanger Centre, Cambridge

Secretariat:
Dr Keith Gibson,
Dr Nanda Rodrigues

Office of Science and Technology:
Dr Eleanor Linton, Ms Christine Gray

TERMS OF REFERENCE

i) To identify, *inter alia*, needs and priorities for technology, information, or communication that would enhance the competitiveness of the UK genome research community, in contributing to fundamental advances and in their application for human welfare and health care, including those that would best be provided as a central resource or by formal networking. To suggest mechanisms for their establishment and management.

ii) To identify the major UK bodies with an interest in genome research, both now and into the future (eg Government, Industry, Charities, Universities, Research Councils, National Health Service).

iii) To document their aspirations and strategies.

iv) To identify the major international players.

v) To document the remits and strategies adopted internationally and to assess cooperation, duplication, omission and successes.

vi) To consider the machinery for regular review of progress within the UK.

EXECUTIVE SUMMARY

The aim of the Human Genome Mapping Project (HGMP) is to identify and characterise all of the estimated 100,000 genes in the human genome and ultimately to sequence the whole genome. This is a formidable task which can only be finished within a 10-15 year period through international collaboration and further developments in technology. The HGMP has enormous potential for the improvement of health and wealth creation. The expected benefits include new diagnostic tools for common diseases and a new biotechnology industry based on genome derived information. We predict that the increased knowledge of basic biology will eventually transform medical practice. Parallel studies in plants and other animals will have an equivalent impact on agriculture and livestock farming.

Internationally, the UK has great strengths in human genetics and genome analysis. Our Report aims to build on these strengths by identifying those areas of expertise where continued investment is required such that the UK maintains and exploits its competitive position. This field is a rapidly moving one and needs to be kept under constant review. The HGMP will directly affect many government departments including Health, the Home Office, the Office of Science and Technology, Health and Safety Executive, Employment Department Group, Trade and Industry, Education and Defence. The partnership between these departments, academia and industry must be nurtured. The setting up of the correct infrastructure now for the analysis and access of all genome information is important for the maintenance of the competitive position of the UK in this field in the 21st century.

We view our recommendations as the minimum required in order for the UK to maintain a competitive position in this field but the financial implications are significant. A means of accessing funds from Government, charities and the private sector needs to be explored, perhaps in a cooperative venture. This Report focuses on the HGMP as the first stage in the understanding of the biology of the human genome which needs specialised funding because it is best carried out in a coordinated fashion. The next stage of studying the detailed function of genes can be funded in the normal peer-reviewed manner, although some attention should be paid now to the development of methods to determine the structure and function of the many novel genes that will arise from the mapping of genomes.

1. GENETIC MAPPING

The genetic map of the human genome is almost complete at a resolution that will permit the mapping of any genetic disorder if the family material is available. The genetic analysis of a disease is the first stage in the identification of the genes involved.

(i) Further work in the UK should be directed towards detailed high resolution mapping of areas of the genome of particular interest, such as those relevant to disease genes.

(ii) The UK is in a strong position to exploit its clinical base and this should form a basis for exploitation in an international setting.

(iii) Priority should be given to the collection of family material in the UK. In particular, the NHS should ensure that the collection of vital information and materials from patients and families is encouraged and enhanced by local providers and purchasers.

(iv) Priority should be given to making facilities available within the scientific community for genetic typing and data analysis; for example, a cost effective way should be developed of distributing typing reagents including PCR primers.

(v) A careful review of the current discussions on the Data Protection Act should be made to ensure that such developments do not pose a serious threat to the effective exploitation of scientific results to national advantage.

2. PHYSICAL MAPPING

A first generation physical map of the human genome is complete although only a small number of the sequenced genes have been placed on the map. With further refinement, this map will be a central resource for genome exploitation and sequencing of the whole genome. The human sequences are maintained in yeast as Yeast Artificial Chromosome (YAC) or in bacteria in vectors such as P1 bacteriophage or cosmids. This allows large quantities of the human sequences to be grown up conveniently for analysis.

(i) Priority should be given to adequate funding of the YAC based physical map construction over a three year period. This should be based on a distributed approach including as many human genetics laboratories as possible.

(ii) There should be a large commitment to map genes, CpG island sequences and other cloned probes onto the physical map. This builds on the UK strength and will greatly facilitate positional cloning projects, particularly the identification of candidate genes in polygenic disease.

(iii) Proposals should be solicited to construct physical maps of large regions at higher resolution than those achieved by the YAC based studies. These should be based upon cosmid or P1 or similar vector systems. This will test the quality of existing maps and is a prerequisite for sequencing the genome.

3. COMPARATIVE MAPPING

The comparative mapping of genomes of several organisms is important for the elucidation of gene function and modelling of human disease. In the future, genome maps of agriculturally important species will form the basis of crop and livestock improvement for both food and non-food applications.

(i) The establishment of a mouse genome centre is urgently required so that advantage can be taken of the strength of the UK in this area. The centre should include expertise in genetic and physical mapping as well as positional cloning. Ideally, the centre should be adjacent to an existing mapping centre to ensure good cross fertilisation of ideas for technology development.

(ii) Priority should be given to the coordination of mapping of cDNAs on the human and mouse maps.

(iii) Strong support for the worm *(Caenorhabditis elegans)*, the fruit fly *(Drosophila melanogaster)*, yeast *(Saccharomyces cerevisiae)* and plant *(Arabidopsis thaliana)* projects should be continued.

(iv) The rat is an important model for pharmaceutical research and is the organism of choice for studying many aspects of whole animal physiology, brain function and polygenic disease. A coordinated programme between the mouse and the rat should be investigated.

(v) The strategic importance of mapping farm animals is very high. Support should be continued for European-wide cooperation to build on existing programmes. The level of support within the UK programme should be investigated.

(vi) Integration, communication and technology transfer between different genome projects is essential. Mapping studies in other organisms with strategic potential such as pufferfish should be monitored and pilot studies initiated where appropriate.

4. DNA SEQUENCING

Ultimately, the DNA sequence of the whole human genome is likely to be determined. However, existing technologies are labour intensive and expensive, and new methodologies are likely to take up to five years to develop. At present, priority is being given to the determination of the sequence of genes. Sequence information on all of the genes in the human genome will eventually have an impact on the understanding of biological processes.

(i) Priority should be given to fund the sequencing of gene-rich regions of the human genome.

(ii) All sequence data should be placed in the public domain.

(iii) Genomic sequencing projects for model organisms with well developed genetics should be supported. This sequence analysis will provide important insights into the biology of living organisms.

(iv) Priority should be given to cDNA, exon and CpG island sequencing in order to ensure that the UK does not lose the more immediate gains to be had by these means.

(v) Large scale DNA sequencing will continue to depend on the method of Sanger for some time. However, it is important to invest in the development of new technologies. Sequencing by hybridisation will become important in diagnostics and its role in primary sequencing should be investigated.

5. GENOME INFORMATICS

The direct product of almost all genome research is large amounts of information, either map positions or DNA sequence. Efficient management of this information is central to the genome project. Both academic and commercial users will be dependent upon ready access to high quality information. This is an area where central policy and resourcing could have an immediate and major impact.

(i) All data should be deposited in centrally funded databases. Access to the databases should be through electronic networks available to both academic and industrial users. The MRC HGMP Resource Centre and the EBI must have very high speed network connections.

(ii) A high quality of database curation is important to make maximal use of data. Support should be made available to UK groups curating subsets of the genomic data. Database development is required and should be supported in the context of these groups.

(iii) Communication between the databases used for different organisms should be coordinated.

(iv) The state of the international central data repositories is of concern. The UK should maintain active involvement with planning for both the sequence libraries (the European copies of which are managed by EBI) and the mapping database, GDB. A strong European component of/counterpart to GDB should be encouraged.

(v) Software development is critical for large scale data collection, and this should be supported in conjunction with large scale sequencing and mapping projects. This software should also be usable by smaller groups.

6. COMMERCIAL OPPORTUNITIES

The commercial opportunities which arise from genome mapping are considerable. Exploitation of the HGMP will not only be important for the biotechnology, pharmaceutical and agricultural industries but also for the health of the nation.

(i) The joint exploitation of genome research by funding agencies should be facilitated. The development of technology transfer from the funding agencies to the public domain should be encouraged. A clear statement from each agency on their strategy for the commercial exploitation of the research they fund is important.

(ii) There is an urgent need to investigate the ways in which the setting up of venture capital funded companies for gene discovery can be facilitated. Such companies should be able to access and exploit genetic and physical mapping technology in the UK or exploit it by another mechanism that leverages all appropriate technology. One possible way is to provide financial incentives for the setting up of small biotechnology companies.

(iii) A LINK-like scheme should be set up which builds on leading edge genome research. This should be implemented so as to involve the best groups in a way that reinforces rather than detracts from their contributions to basic knowledge.

7. PUBLIC EDUCATION AND TRAINING

The medical opportunities that arise from the HGMP will not benefit the health of the nation unless there is an infrastructure in which the data can be interpreted and exploited.

(i) Models which allow the commercial development of large-scale genetic testing need to be explored urgently.

(ii) Education programmes in genetics should be set up at all levels from primary school through to postgraduate medicine in order to maximise the health impact of the HGMP.

(iii) There is a requirement for a careful assessment of the needs of industry, government departments and academia for individuals with training in genome analysis and its applications.

1. INTRODUCTION

1.1 BACKGROUND

The main objective of the Human Genome Mapping Project (HGMP) is to identify and characterise all of the estimated 100,000 genes and ultimately to understand the whole genome. The first stage is to construct a gene map and then to sequence the whole genome. The immediate benefits will be in health care through the identification and sequencing of genes that confer susceptibility to common diseases such as diabetes, heart disease and psychiatric conditions. These diseases are a major cause of morbidity in our society and are a major financial burden on the NHS. In the longer term, improved diagnostics will be replaced by improved treatments.

The determination of the sequence of 3,000 million base pairs which constitute the human genome is an enormous task requiring both international collaboration and improvements in technology in order for it to be completed within a time-scale of 10-15 years. Using current technology, the total budget for this endeavour would be approximately £2 billion. This figure is likely to be substantially reduced by the introduction of novel approaches to DNA sequencing.

Although the UK budget for genome mapping has been modest, the strong British tradition in human genetics has enabled the UK to make significant contributions to the international human genome mapping effort. UK generated resources such as yeast artificial chromosome (YAC) libraries are being used by the international community and the genetic map of the mouse genome, generated by a European Consortium, is being used as the backbone towards a complete genetic map of the mouse. The UK MRC Human Genome Mapping Project was the first to establish a successful Resource Centre which is now being emulated in other countries. The UK is also particularly strong in human genetics research and thus in a relatively good position to take advantage of the results of the HGMP.

One of the important features of the HGMP is the extraordinarily rapid growth in information and the equally rapid growth in technological innovation. To ensure that the UK continues to make a significant contribution to the HGMP and maintains a position whereby it can readily build on the results of the HGMP, the Office of Science and Technology established the Advisory Committee on Human Genome Research with membership from the Research Councils, Academia, Industry, Charities and Government to advise them on the Project. The membership of the Advisory Committee is at Annex 1. The Advisory Committee in turn appointed a group of experts actively engaged in genome mapping activities to advise them. The membership and terms of reference of the Expert Working Group are on pages 4 and 5.

1.2 SCOPE OF THE REPORT

The major thrust and focus of the Report is on the Human Genome Mapping Project. Mapping studies in other genomes are pertinent and in some cases very important to the progress of the human studies. We have therefore examined work that is both directly and highly relevant to the HGMP. We have, however, made no attempt to review or comment in detail on other genome mapping projects.

The ethical and social implications of the HGMP have also not been reviewed in this Report as they are being dealt with by other bodies. The Nuffield Council on Bioethics had commissioned a Report in this area which was published in December 1993 (Title: Genetic Screening - Ethical Issues).

In view of the rapid progress in this field, the preparation time of this Report has been kept to a minimum with only very selective additional consultation and advice from outside the membership. Colleagues in Industry, Academia, Research Councils, and Charities were consulted on a number of key questions. A list of those consulted is at Annex 2. The questionnaire sent out is at Annex 3.

Most of the present activity in the HGMP is focused on academic institutions although several commercial groups are now setting up their own in-house research programmes. A major challenge is to build on British academic success by promoting interactions with the commercial sector. The bulk of the discussion focuses on the scope of the basic work being done in the UK, how interactions with the commercial sector can be promoted and how this fits into the international effort.

During the course of the preparation of this Report, several other reviews have been undertaken. In the UK, the Government has issued its White paper 'Realising our potential - A strategy for Science, Engineering and Technology' (London, 1993, HMSO, Cm 2250) and the Medical Research Council (MRC) has undertaken a major review of its Human Genome Mapping Project. Major assessment of National Human Genome Mapping Programmes is also being undertaken in Canada and USA. The recommendations of these reviews have been taken into account when formulating the conclusions of this Report.

2. CURRENT SUPPORT FOR HUMAN GENOME MAPPING

There have been several excellent recent reviews of the international efforts in human genome mapping (see Annex 4 for list of references). Rather than duplicate this effort, this Report summarises the conclusions of these papers briefly as background information for the future recommendations.

2.1. HUMAN GENOME RESEARCH IN THE UK

Human genome research in the UK is funded through the Government and the Charities such as the Wellcome Trust and Imperial Cancer Research Fund. Funding exists in specific genome centres as well as in research institutes and university departments.

2.1.1. Medical Research Council (MRC)

The direct government support for human genome research has been channelled through the Medical Research Council (MRC). In 1989, the MRC was allocated £11m over a three year period to set up the UK Human Genome Mapping Project (HGMP). This support constituted new, 'ring-fenced' money. In April 1992, a sum of £4.5m per annum was consolidated into the MRC's Grant-in-Aid and allocated specifically for genome research. Genome mapping projects were reviewed by a Directed Programme Committee and were not assessed by external review. This enabled projects to be considered expeditiously in keeping with the rapid advances in the field. Following the recent review of the UK HGMP, ring-fenced funding for genome research has been removed. Applications for genome research in the future will be considered together with other bids for MRC resources and all applications for support will be peer reviewed. An overarching HGMP Coordinating Committee has been set up to develop policy and to identify strategic scientific areas within the MRC Programme. This Committee includes representatives nominated by interested Charities and other Research Councils and therefore should be able to facilitate the coordination of the UK HGMP. In addition, an Industrial Advisory Group has been established which includes representatives nominated by interested private sector companies, which will feed into the work of the HGMP Coordinating Committee. This Group will also feed into the work of the MRC Gene Therapy Coordinating Committee and the Genetics Approach to Human Health Steering Committee.

The new arrangements for funding highly rated scientific projects recently introduced by the MRC should allow rapid assessment of applications in the area of the HGMP. Removal of ring-fencing means that applicants submitting proposals for genome research can now bid against the whole of the Council's research budget rather than the smaller ring-fenced budget originally allocated for the HGMP.

New funding has been made available by the MRC during 1993 to promote work on comparative mapping. The MRC allocated £1.5m to this initiative over a one year period for biological and computing studies.

The MRC HGMP funds also support a Resource Centre. The MRC HGMP Resource Centre was founded at the Clinical Research Centre at Northwick Park Hospital and was not adjacent to an academic group involved in genome mapping. It is a neutral and independent Centre which has become a national facility providing services and reagents to the entire UK scientific community. The MRC HGMP Resource Centre acts as a national repository and site for systematic programmes of data generation as well as a distribution and reference centre for human and mouse mapping resources. It provides a communal service which alleviates the burden of repetitive and expensive tasks from individual scientists and acts as a focus for sustained and systematic programmes. In addition, the Resource Centre also provides computing facilities and access to international databases and computer training courses. There are currently about 2000 registered users of the Resource Centre at a cost of just over £1m per annum to the HGMP. The services offered by the Resource Centre are free to academic users in return for data deposition. Industry may use the services for a small charge to cover costs. The facilities are already being used by Glaxo, SmithKline Beecham, British Biotechnology, Yamanouchi Research Institute, Zeneca Pharmaceuticals and the Wellcome Foundation.

The MRC HGMP Resource Centre has held a number of open days for industry and for agencies such as the Charities, other Research Councils, the Department of Trade and Industry and the Patent Office.

The MRC HGMP Resource Centre will be moving to Hinxton Hall adjacent to the MRC/Wellcome funded Sanger Centre in 1994 thus placing it in proximity to academic activities relevant to genome mapping. Changes in management and direction resulting from this relocation should enable the Resource Centre not only to build on existing resources, but also to establish programmes at the leading edge of genome research fully integrated with genome programmes elsewhere in the UK.

In recognition of the importance of exploiting information generated by the HGMP, the MRC has recently launched the Genetic Approach to Human Health (GAHH) initiative. This initiative seeks to promote gene identification, investigation of gene function and the practical exploitation of the new information in areas such as gene therapy and diagnostics.

HGMP funding has been used to fund studentships, fellowships and travel awards which enable researchers to learn new technologies. The MRC HGMP has funded the costs of organising Single Chromosome Workshops in the UK and has supported UK researchers to attend such workshops abroad. The annual 'Users Meetings', also funded by the MRC, have uniquely brought together members of the UK genome community on a regular basis, initially to inform them about national and international administrative issues, and latterly to learn about new scientific developments.

In addition to directly funding the UK HGMP, the Government also provides funding indirectly through the European Commission (EC). The European Community Human Genome Analysis Programme (1990-1992) is one component in the portfolio of European genetics-centred work. Various Community countries have national programmes, and the aim of the EC programme has been to ensure a significant European contribution to the world-wide effort to map the human genome. The MRC acted on behalf of the Department of Education and Science as the UK lead for the EC Human Genome Analysis Programme (1990-1992). Responsibilities of the lead organisation include the dissemination of information about the Programme throughout the UK and UK representation on the Programme Management Committee.

Support for genome mapping activities will be continued in the Third Framework Programme as part of the Biomedical and Health Programme BIOMED I (1990-1994) as Area III (Human Genome Analysis). However, the bureaucratic procedures involved in applying for funding have been a cause of grief for many scientists. The forms are very long and detailed and are released only weeks before the deadline for submission. This may have resulted in the loss of opportunities for real collaboration in

Europe. Genome research will form part of the BIOMED programme under the fourth EC Framework. This will run from 1994-99.

2.1.2. *Agricultural and Food Research Council (AFRC)*

The Agricultural and Food Research Council (AFRC; from April 1994 the AFRC will be combined with some components of the Science and Engineering Research Council to form a new Research Council: The Biotechnology and Biological Sciences Research Council) has a long established portfolio of research on genome mapping and analysis of various species. A major aim is to identify genes controlling commercially important traits such as growth, fertility and disease resistance. There are strong links and close collaboration with the MRC in the area of genome analysis with coordination of comparative mapping programmes. In animals, the AFRC supports research on pig mapping as part of PiGMap, an European initiative funded under the EC BRIDGE programme and in part from the Science Budget. There is also limited funding for work within the AFRC on poultry and cattle funded by the Science Budget or by the Ministry of Agriculture, Fisheries and Food (MAFF) respectively. The chicken genome mapping project within the AFRC has produced the first linkage map of the chicken genome with 150 marker loci. This map is now the centre of a number of international collaborations which are rapidly developing the map and applying it to analyse traits. The AFRC's new coordinated programme in genome analysis will build on this established base concentrating on pigs, chickens and cattle. Additional funding has already been made available for sequencing equipment at the Roslin Institute and the Institute for Animal Health.

In plant research, the AFRC is one of the groups leading an international programme to map the genome of the model plant *Arabidopsis thaliana* (Institute of Plant Science Research). The AFRC also has active projects mapping the genomes of pea, wheat, barley, grasses (Institute for Grassland and Environmental Research) and oilseed rape (Horticulture Research International). The commercial importance of these projects has been recognised by considerable industrial investment.

2.1.3. *Wellcome Trust*

The Wellcome Trust has committed substantial funding to the establishment of two key centres in the UK involved in genome research. The Sanger Centre in Cambridge involves a major sequencing and physical mapping initiative of the human genome, as well as continued sequencing of the worm *Caenorhabditis elegans* (a project sponsored through the MRC) and of yeast. This Centre will be the major focus together with the MRC HGMP Resource Centre and the European Bioinformatics Institute (EBI), for the HGMP in the UK. The EBI is an outstation of the European Molecular Biology Laboratory (EMBL). The capital costs of the EBI in the UK have been provided by the Wellcome Trust and the MRC.

In Oxford, another Wellcome funded major initiative of £13m over five years aims to study the genes involved in complex diseases. These are diseases in which multiple genes interact with environmental effects to produce a clinical phenotype. Examples are diabetes, hypertension and asthma.

In recognition of the importance of genetics in medical research, the Wellcome Trust has set up a Genetics Interest Group to oversee applications in the general area of human genome mapping, human genetics and gene therapy.

2.1.4. Imperial Cancer Research Fund (ICRF)

The Imperial Cancer Research Fund (ICRF) has made significant contributions to the HGMP because of its relevance in the understanding and treatment of cancer. Scientists at the ICRF have generated important resources which have been used internationally. These include large insert YAC libraries of the human, mouse and yeast genomes; chromosome-specific cosmid libraries from flow sorted chromosomes; cDNA libraries; P1 libraries, somatic cell hybrids. A high throughput of information has been achieved by the screening of high density filter grids. Scientists at the ICRF have also been responsible for establishing the DNA probe bank in the UK with the help of a grant from the MRC. The probe bank now constitutes an important resource at the MRC HGMP Resource Centre. ICRF Laboratories at Clare Hall have been involved in the setting up of somatic cell hybrid mapping resources for the localisation of sequences on chromosomes and selected primer pairs for genetic mapping.

The ICRF also encourages development of sequencing technology and has undertaken large sequencing projects. Scientists have been developing high throughput techniques for data generation, made possible by a close integration of experimental techniques, automation and informatics, together with the development of new concepts on data acquisition, data storage and data analysis.

2.1.5. Other Charities

Many charities other than those noted above are making important contributions to the HGMP. Their support for human genetic studies is invaluable through the collection of clinical material and funding of research on specific diseases. The continued support of these charities is important for the future progress of the HGMP. A list of charities contributing actively into human genetics can be obtained from the Association of Medical Research Charities.

2.2 HUMAN GENOME RESEARCH ELSEWHERE IN EUROPE

European initiatives in human genome research have been developing although not on the same scale as the USA. The combined contributions of two genome centres - Genethon, a £6.5 million automated laboratory facility supported by the French Muscular Dystrophy Association and the Centre d'Etude Polymorphisme Humain (CEPH) - have put France in second place after the USA in the amount of human genome research that is being carried out. In 1992 Genethon had already finished a map covering nearly 28% of the genome and almost the entire chromosome 21. It has also generated a valuable resource of nearly 2000 microsatellite markers for the genetic map. They published a first generation map covering 90% of the genome in 1993. Many of the European countries are engaged in classical genetics research and research on genetic diseases which underpins the international human genome research initiative. The funding for such studies is largely through existing mechanisms of support. The countries that have a national genome programme for which earmarked funding has been committed are Denmark, France, Germany, Italy, Netherlands, UK and USSR. The Nordic countries have proposed a cooperative genome initiative which would allow them to contribute jointly to the human genome mapping effort.

The EC Human Genome Analysis Programme had a budget of £11.5m over a period of two years (1990-1992). The programme was then subsumed into the Biomedicine and Health Programme (BIOMED 1, Area III). The topics encompassed by the programme include: Improvement of the human genetic map; Physical mapping; Improvement of methodology; Data handling; Training; Ethical, social and legal issues. The various genome analysis programmes sponsored by the European Commission offer smaller European countries, European Free Trade Association (EFTA) and European Cooperation in the field of Scientific and Technical research (COST) countries a means of securing additional support for genome research. There should be complementarity between these funds and national funds so that total resources can be optimally used to European advantage. Eastern European countries seem not to figure large in genome research at present.

2.3 HUMAN GENOME RESEARCH IN THE USA

The HGMP was initiated in the mid-nineteen eighties. The first public discussion of the project was at a symposium on the 'Molecular Biology of *homo sapiens*' in Cold Spring Harbor in 1986. Funds amounting to a sum of £18.8m and £12.3m were made available by the USA National Institutes of Health (NIH) and the Department of Energy (DOE) respectively during the period of 1988-89. The operating budget for 1992 was £69.8m from NIH and £40.2m from the DOE. The estimated budget for 1993 was £70.8m from NIH and £43m from the DOE.

The HGMP's five year goals (October 1990-September 1995) included completion of a fully connected human genetic map with markers spaced an average of 2 to 5 centimorgans apart; assembling STS maps of all human chromosomes with markers spaced at approximately 100,000 base pair intervals; generation of overlapping sets of cloned DNA or closely spaced unambiguously ordered markers with continuity over lengths of 2 million base pairs for large parts of the human genome; improving current and developing new methods for DNA sequencing that will allow large scale sequencing of DNA at a cost of 50 cents per base pair and determining the sequence of an aggregate of 10 million base pairs of human DNA in large, continuous stretches in the course of technology development and validation.

The NIH National Centre for Human Genome Research (NCHGR) held a meeting in Hunt Valley, Maryland in 1993 as part of its process for developing a research plan for the next phase of the HGMP. The meeting was intended to (1) appraise project accomplishments likely to be achieved by the end of the initial five year phase; (2) generate creative and novel ideas for building on this progress to meet long-range goals; and (3) consider ways in which the project could most effectively and broadly benefit present and future research in biology and medicine. The successful development of new technology has been essential for the HGMP and will continue to be critical in the future. It was decided that although the HGMP must continue to pursue its original goal of obtaining the complete human DNA sequence, it was necessary to ensure that technologies were developed that would allow the full interpretation of the DNA sequence once it was available. In order to increase emphasis on this area, an explicit goal related to gene identification was added. This recommendation is similar to the cDNA strategy (gene map) favoured by Europe and Japan. Another important recommendation was related to transfer of technology both into and out of centres of genome research, which in future would be encouraged and enhanced.

2.4 HUMAN GENOME RESEARCH ELSEWHERE

Outside Europe and the USA, Japan is the only other country to have a major human genome programme with a budget of £20m in 1992, although there are other countries embarking on smaller programmes. In Japan, the Ministry of Education, Science and Culture is the major funding body for the genome programme which consists of five activities; grant-in-aid for human genome research, a central human genome centre, five local genome research centres, international symposia and workshops, special fellowship programme and grant-in-aid for genome informatics. The Science and Technology Agency and the Ministry of Health and Welfare also provide funding for genome research. Japan has concentrated on mapping human chromosomes and also invested substantially in a cDNA strategy which they refer to as a 'body mapping' strategy. Other countries with smaller human genome programmes include CIS (ex-Soviet Union), Canada, South Korea, New Zealand and Australia. Alongside national programmes there are international agencies that are involved in human genome research. These are: the Commission of the European Community (EC), the Human Genome Organisation (HUGO), the European Molecular Biology Organisation (EMBO) and the United Nations Education Science and Culture Organisation (UNESCO).

2.5 INTERNATIONAL COORDINATION OF HUMAN GENOME RESEARCH-HUGO

The Human Genome Organisation (HUGO) was founded by scientists in 1988 at the Cold Spring Harbor Meeting on Genome Mapping and Sequencing to promote the international coordination of human genome research.

Until 1990, the integration of the human genome information had been carried out through meetings of scientists at Human Genome Mapping Workshops held approximately every two years. Since then, because of the large volume of data, this role has been taken over by individual chromosome workshops which have been held annually for all the individual chromosomes. The coordinating role of HUGO has been crucial in bringing together into a global initiative the mapping and sequencing work being done in many different countries. HUGO's priority during the last two years has been to develop a programme of international Single Chromosome Workshops. These have the specific function of collecting, sharing and distributing new data. They also facilitate the building of consensus genetic and physical maps and the spreading of information and technology. The annual Chromosome Coordinating Meetings, where a few representatives of each of the chromosomes meet, are also held under the auspices of HUGO. Human Genome Mapping Workshops sponsored by HUGO, where the whole genome community meet are held biennially. HUGO has worked with the genome community to develop guidelines for all aspects of the running of the workshops to ensure the uniform handling of data and entry into GDB and has also negotiated financial support. As an adjunct to the coordination of scientific data, HUGO is collating information about the availability of biological resources and has also implemented a short-term travel award programme to provide for rapid updating of laboratory expertise. HUGO has initiated informed international discussion about the ethical, legal and social issues arising from the HGMP and is also playing a valuable role in facilitating the flow of information about genome matters throughout the genome community.

3. RECOMMENDATIONS FOR FUTURE DEVELOPMENTS

The Expert Working Group examined each of the main areas of human genome mapping research and their exploitation, identifying strengths in the UK and evaluating potential areas for investment in the future.

3.1 GENETIC MAPPING

The genetic map of the human genome is well advanced due to the efforts at Genethon and several Genome Centres in the USA (see Sections 2.2 and 2.3). There are still gaps to fill in and more detailed maps will be required for areas of particular interest.

3.1.1. Generation of new markers

Because of the efforts elsewhere to generate new microsatellite markers, there seems little merit in developing a programme of marker generation in the UK. The markers being generated by the Genethon group are now widely available. The UK at present does not have a competitive programme, nor would it be prudent to put resources into such a programme when the development of large numbers of markers is so well advanced elsewhere. However, the UK has produced markers for the mouse genome (see Section 3.3).

3.1.2. Family collection

The structure of the National Health Service (NHS) in this country provides one of the best opportunities to develop a lead in the study of genetic disease. Large family collections of both rare disorders and common multifactorial disease can be readily collected at present through the NHS. This contrasts with the situation in the USA where the majority of the health care is in the private sector, and access to good patient material is extremely difficult. Another beneficial factor in the UK is that the population is relatively static compared to the USA, and collection of families is easier. Priority should be given to the collection of pedigrees in this country. These activities are very expensive and time consuming, but underlie the whole programme in human disease genetic mapping. Family collections are most successful when they are driven by a group with a specific interest in the genetics of a single disease. Such groups can be a focus for the collection of families which are then deposited for wider access and EBV transformation in Porton Down. We would strongly recommend that all major family collections are made available to the genetics community at large. EBV transformation is essential to ensure that the patient material is a renewable resource and can be used by multiple groups in the future. There are already programmes established for the collection of multiplex families suffering from the common multifactorial diseases. Many of these need additional support and this should be encouraged, provided that the family material is made widely available in the UK.

A major concern in the collection of family material is the reorganisation of the NHS. The funding implications for GP practices and hospital trusts need to be addressed directly. We recommend close cooperation between the NHS, relevant charities and the MRC to ensure that the collection of family material is successful. There is increasing difficulty in obtaining information and samples from extended families, as hospitals and GP fund-holders become concerned to identify and restrict activities which fall outside their contractual obligations. A clarification of the obligation on medical practitioners to assist in reasonable research would be helpful.

Under the Data Protection Act, research use of computerised information relating to identifiable individuals is only exempt from the Act if it has been fairly obtained; that is now being interpreted as meaning fully informed consent to its use - and that includes current use, not just that for which it was collected. An EC Directive currently being drafted could well impose further restrictions. These developments pose a serious potential threat to the effective exploitation of scientific results to national advantage.

The ethical implications of genetic testing are not covered in the terms of reference of the Expert Working Group. The reader is referred to the recent Report from the Nuffield Council on Bioethics for further information in this area.

3.1.3. Marker testing

a) Single gene disorders

Although the majority of diseases caused by single gene defects have been well studied, there will be a continued need for mapping genes which cause comparatively rare diseases. Study of these diseases at the genetic level, will provide new insight into the pathophysiology of the diseases as well as providing reagents that can be used for prenatal diagnosis. Many such family collections are held by groups outside the main centres for human genetics. The proliferation of automated sequencing machines and the availability of a large set of fluorescently labelled markers should permit many such family collections to be studied in one of the centres for human genetics which have automated machines. Such a distributed approach to genetic mapping would need to be facilitated by a central resource of fluorescently labelled markers that could be made available for such mapping endeavours. This central resource of markers should be strongly encouraged and perhaps held by the MRC HGMP Resource Centre. Individuals with family collections in centres with automated sequencers could then simply use the markers to test their families. For groups with good family collections but no access to sequencing machines it might be possible, through the MRC HGMP Resource Centre, to have a few machines which would be made available for individuals to come to the Resource Centre and test their family material. The size of such a facility would be limited by the number of groups with families who applied to use the service and by the amount of additional expertise that was available. Statistical advice for data analysis could also be provided through the Resource Centre by electronic mail.

b) Marker typing in polygenic disease

The largest impact of the new mapping technologies is likely to be in the analysis of genetic susceptibility in complex diseases. Many diseases are the result of the interaction of environmental factors (multifactorial), such as exposure to infectious agents, with several different genes. Defining the susceptibility genes in these polygenic diseases will address some of the most common diseases to afflict our society.

There is already one major centre funded to characterise genetic susceptibility in polygenic disease in Oxford. This programme can provide the necessary statistical back-up and large scale mapping facilities to allow good family collections to be characterised rapidly for genetic susceptibility. Other centres have specific interests in a single polygenic disease. Again, these centres would benefit from a central resource of fluorescently labelled primers for marker typing. These centres which have collected substantial amounts of family material for an individual disease and which have substantial genetic expertise in other areas, should be encouraged to establish their own typing programmes.

3.1.4. Financial implications

The collection of family material is expensive. It costs approximately £100 to collect and store each patient sample. In many cases it will be necessary to supplement existing family collection programmes in diseases like Type II diabetes, psoriasis, inflammatory bowel disease, osteoarthritis etc. We estimate that an extra investment of between £1m to £2m will be required over the next five years to adequately support family collection activities.

A central supply of fluorescently labelled primers for microsatellite typing on automated sequencers would have many advantages. The common set of markers would enable different data sets to be easily integrated and development costs would be greatly reduced. Added to this are the economies of scale gained from a single large scale synthesis. An investment of £1m would provide 1000 markers in sufficient quantity to supply the whole UK effort for several years.

3.1.5 Recommendations

(i) Further work in the UK should be directed towards detailed high resolution mapping of areas of the genome of particular interest, such as those relevant to disease genes.

(ii) The UK is in a strong position to exploit its clinical base and this should form a basis for exploitation in an international setting.

(iii) Priority should be given to the collection of family material in the UK. In particular, the NHS should ensure that the collection of vital information and materials from patients and families is encouraged and enhanced by local providers and purchasers.

(iv) Priority should be given to making facilities available within the scientific community for genetic typing and data analysis; for example, a cost effective way should be developed of distributing typing reagents including PCR primers.

(v) A careful review of the current discussions on the Data Protection Act should be made to ensure that such developments do not pose a serious threat to the effective exploitation of scientific results to national advantage.

3.2. PHYSICAL MAPPING

The HGMP has the final goal of producing a detailed description of all human genes and their controlling regions. This will revolutionise the health care industry. Ultimately the description will be at the sequence level but there is a general recognition that systematic sequencing of the entire genome is not the best use of resources at this stage. The construction of a hierarchical set of integrated physical maps of the genome has been identified as the next logical step prior to a full sequence-based analysis. This includes the placing of genes on to the map as their sequences become available.

National and international efforts have become centred on the construction of physical maps at the level of overlapping cloned DNA fragments carried by YAC or cosmid (or other moderate capacity) vector systems. It is unlikely that new technical developments will alter this focus and it is reasonable to expect the methods of analysis to be moderately stable, and the map will be completed in the next two to three years.

Three methods are being employed to construct overlapping cloned DNA maps. STS mapping uses DNA sequences that are amplified by PCR. Overlapping clones are identified by showing they contain the same STS. Hybridisation analysis involves identification of overlapping clones by hybridisation screening of arrays of YAC or cosmid clones immobilised on filters. Overlapping clones are identified since they hybridise with the same probe(s). Fingerprint analysis is carried out by generating a pattern of DNA fragments from individual YACs or cosmid clones that can be analysed by gel electrophoresis. Overlapping clones can be identified since they will have (partially) similar fragment patterns.

Data collection is almost invariably by computerised analysis of digital images and is closely integrated with database systems for storing and manipulating information. It is anticipated that massive data stores (>70 Gbytes) will be required for the UK programme alone.

3.2.1 International activities

France, through the unique activities of the Genethon/CEPH collaboration, has made significant progress on construction of YAC based maps; the STS generated map for chromosome 21 has been published and also a fingerprint based method for the entire genome. Most recently, they have presented a first generation map of the whole genome based upon multiple analyses of their YAC libraries; while incomplete and no doubt subject to errors, this is a very considerable achievement that will provide a framework map upon which all genome projects can build. The CEPH libraries have been incorporated into the UK effort.

The USA, under the NIH, has refocused its efforts from a multicentre, chromosome based STS analysis to a single approach at the Whitehead Institute in Boston. Single chromosome genome centres (chromosomes 4, 7, 11 and 22 *inter alia*) are still funded. Under the DOE, cosmid and YAC based analyses of chromosome 16 and 19 are well advanced.

Outside of these two countries there would appear to be no other significant physical mapping initiatives. Japan has focused on regional mapping and cDNA sequencing in particular.

3.2.2. Physical mapping in the UK

The MRC HGMP has initiated a systematic project to construct a YAC based physical map of the human genome using hybridisation analysis. The project is explicitly structured with two major goals; to assemble an overlapping YAC map of the entire genome and, secondly, to integrate into this essentially anonymous physical map the position of all the DNA markers that define the existing body of genetic, biological and physical data that relates to the human genome. This strategy has been explicitly formulated to incorporate directly to primary map construction where other maps are lacking, or deepening the coverage of regions already contained in YAC clones. The second goal is of the highest priority for those regions where map construction and YAC coverage is complete; without these analyses, the YAC map does not contain any relevant biological information and is therefore useless to clinical, genetic or industrial researchers.

Underlying this centralised approach is a highly distributed level of analysis that will focus on specific chromosomal regions. This level is of great practical importance since it draws on the reagents of participating laboratories who have large collections of probe DNA molecules (genes, cDNAs, chromosomal rearrangements, cell lines, patient materials) isolated from small regions of the genome. It is also the most effective method for ensuring that medical researchers have direct access to genome mapping data at a very early stage, well in advance of completion of a coherent genome wide map. This ensures that the project will contribute to clinically relevant research at the earliest stages.

YAC map construction

The project is centred upon the construction of the YAC based map of the human genome within a two to three year time frame using 10,000 hybridisation analyses of YAC filter arrays. These will include the CEPH YAC clones, ensuring that the data generated can be applied both to our own expanding map and to extending that of CEPH. YAC filters are generated at the ICRF, Sanger Centre and the MRC HGMP Resource Centre and will include systematic genome wide screening. The 10,000 probes will consist of anonymous fragments, cDNA clones and CpG island fragments. The latter probes are particularly promising since they represent a general class of sequence that identifies the position of a gene nearby. The UK has pioneered the construction of CpG island libraries and this is an unique resource. Choice of probe is of primary importance to the second goal of the project, that of map integration.

Map integration

The properties, in the broadest sense, of the human genome are defined by specific DNA fragments that can be recognised genetically. These may encode genes, recombination markers, or breakpoints, or define intervals or fragile sites, or contain specific types of DNA sequence, including repeats and micro- and mini-satellites: all can be defined by 'marker' DNA probe molecules. These markers will necessarily define many of the biological and medical features of the chromosomes. Map integration can be reduced to the identification of the position of marker DNAs within the developing cloned DNA map. The appropriate choice of markers for hybridisation analysis will therefore be of central importance to the project.

Distributed analyses

Particular regions of the genome, large regions or whole arms of chromosomes, have been subjected to very detailed analyses by many groups: this has produced markers at very high local density and very detailed genetic information. It would be experimentally impossible for the whole genome to be analysed centrally with such detail and so the project is being distributed between different laboratories to incorporate this rich body of data.

Collaborations are envisaged from groups who have substantial collections of probes from restricted regions of the genome. Filters are distributed by Dr Hans Lehrach (ICRF) and by the MRC HGMP Resource Centre to these groups, who will be contractually obligated to probe them with substantial numbers of regionally defined clones.

This programme will occur in parallel with the distribution of filters to any requesting group who wish to isolate a YAC clone of a single gene and the MRC HGMP Resource Centre also provides a PCR screening facility for this purpose. These activities facilitate the process of specific, normally clinically defined, research project goals and also contribute to map integration, since it is a condition of use of the filters that information is returned to the distributing centre as to the identity of the appropriate YAC clone(s).

Map verification

The construction of an internally consistent YAC map of the genome does not, for complex reasons, guarantee that the map is necessarily 'correct' in every detail. Verification of the map is an important and difficult proposition and is probably best carried out by the end user of the map. The collaborating centres are ideally positioned to have a major input into this area and this is a second vital feature of this level of involvement within the programme.

3.2.3 Physical mapping and DNA sequencing: lower level clone map analyses.

The construction of a contiguous YAC based physical map of the genome is necessary but not sufficient to allow the project to develop into its next phase - the elucidation of the DNA sequence of large genomic intervals. The reason for this is primarily technical- the generation of DNA sequences directly from YAC clones is still relatively more difficult than the use of cosmid clones as sequencing templates and this will remain true for the foreseeable future. It is also clear, from recent papers reporting the successful positional cloning of a number of disease causing genes, that cosmid sized DNA fragments (35kb) are the preferred substrate for gene searches by physical techniques.

Both of these observations highlight the importance of studies constructing clone maps directly at the cosmid and P1 level. We still have a poor understanding of the constraints on the representation of DNA sequences in genomic libraries constructed with any cloning vector: it is generally accepted that all vector/host systems (including YACs) have difficulties in carrying some DNA sequences. It follows that it is desirable to construct primary cosmid or P1 libraries directly from genomic DNAs rather than from YAC contigs. Integration of the two sorts of map (YAC and P1/cosmid) is possible and technically important for full genomic representation.

In the UK, projects on fingerprint based cosmid maps of the Y and 11 chromosomes have been initiated and hybridisation based analyses of cosmid and P1 libraries from individual, flow sorted chromosomes (X, 21, 22) are in progress. There are also a number of regionally limited projects on constructing cloned DNA maps of parts of the antibody gene family and the major histocompatibility locus (on chromosome 6). These sorts of analyses are critical to the development of a successful genome project and should be encouraged.

3.2.4 *Unique UK strengths*

The UK has a unique strength in the production and distribution of arrays of cloned DNAs immobilised on filters. The robotics to support this have been developed primarily by Dr Hans Lehrach (ICRF) and Dr David Bentley (Sanger Centre). None of our competitors can match either the throughput or the density of clones on the filters and these robotic facilities are central to our efforts. The MRC HGMP Resource Centre can play a unique role in the distribution of resources, provided it is more closely integrated into the project. This should be a key feature of the future shape of the programme but the Resource Centre will need expansion and will have to attract a high calibre of scientist if it is to assume its central role. Both conditions require additional support.

The distributed nature of the UK genome mapping effort will generate a YAC map that is fully integrated into the existing genetic and physical map of the genome. It is not clear that any of the competing international activities can do this except where specific centres are carrying out similar actions. Incorporation of YACs from the same sources as the USA genome centres ensures that data integration will be technically possible between different national efforts.

The UK has extensive experience in the cosmid level analysis of large genomes. This is important for the next phase of the project - the establishment of regional DNA sequence.

3.2.5 *Resources and project organisation*

The MRC programme is the only coordinated national programme of genome research but this organisation does not control all of the financial input into the project. As it stands, it cannot fund all of the priority areas. The current MRC investment in the pilot project is about £296,000 which is sufficient only for limited (and successful) studies at the ICRF and the Sanger Centre. These suggest that a five man group can carry out 64 hybridisations per week, allowing project completion within just over three years. Such a group costs about £300,000 pa and if the project is to remain internationally competitive then completion must be within two years. This argues that an investment of some £600,000 pa is required for central activities. No systematic programme of identifying the 'distributed' collaborators has been initiated and funds for this are limited: a properly detailed programme and a call for participants should be coordinated through the funding organisations. The problem of data storage is acute, although the MRC has recently started to address this: an investment of £100,000 has been made in computer storage facilities at the MRC HGMP Resource Centre.

The role of the MRC HGMP Resource Centre is not entirely clear at present. It needs to occupy a more central role within the national programme and is the logical organisation for handling much of the filter production, distribution and data storage. This has been discussed above.

Within the scientific programmes currently funded in the UK there is limited work on higher resolution mapping (with cosmid or P1 etc libraries) and since these represent the substrate for future detailed studies, they should be encouraged.

The sharp focus of the project being developed under the MRC initiative is without doubt the correct scientific way forward: however, good coordination between the participating groups and funding agencies is vital if the project is to remain internationally competitive.

3.2.6 *Recommendations*

(i) Priority should be given to adequate funding of the YAC based physical map construction over a three year period. This should be based on a distributed approach including as many human genetics laboratories as possible.

(ii) There should be a large commitment to map genes, CpG island sequences and other cloned probes onto the physical map. This builds on the UK strength and will greatly facilitate positional cloning projects, particularly the identification of candidate genes in polygenic disease.

(iii) Proposals should be solicited to construct physical maps of large regions at higher resolution than those achieved by the YAC based studies. These should be based upon cosmid or P1 or similar vector systems. This will test the quality of existing maps and is a prerequisite for sequencing the genome.

3.3 COMPARATIVE MAPPING AND MODEL ORGANISMS

To facilitate our understanding of human biology and disease it is essential to study a diverse range of model organisms whose genomes are less complicated. In spite of the differences many genes fulfil similar basic functions. Hence our knowledge of gene function and biological mechanism is underpinned by studies of yeast *(Saccharomyces cerevisiae)*, plants *(Arabidopsis thaliana)*, the fruit fly *(Drosophila melanogaster)* and the worm *(Caenorhabditis elegans)*. Strategies for mapping, sequencing and informatics are best developed first in these organisms because of their lower genetic complexity and ease of manipulation. In addition, the analysis of model systems has an intrinsic importance both in understanding the organisms themselves and to industry and agriculture.

The direct and relevant models for human biology and genetics have to be mammalian systems and of these the most highly developed as a research tool is the mouse. However, to gain the full benefits of comparative genome mapping several other strategically important mammalian species must also be studied.

3.3.1 *The mouse*

Of all the mammalian species the mouse has the most highly developed genetics; its size and short generation time have allowed large scale mutagenesis programmes and extensive genetic crosses. The UK has had a distinguished history in mouse genetics. Another key advantage of using the mouse is the ability to manipulate its genome. It is possible to introduce DNA fragments into the germline to study the function and expression of genes. It is also possible to target mutations to specific genes using homologous recombination in Embryonic Stem cells. A very recent advance has been the ability to introduce very large fragments of DNA cloned on YACs into the mouse genome in a functional form. This very powerful new assay system is amenable to human as well as mouse genes.

3.3.2 *Benefits to man of mouse genetics and comparative genome analysis*

There are several ways in which an intensive mapping effort in the mouse and comparative mouse/ human maps will pay dividends for human genome analysis.

1. Knowledge of the mouse map will facilitate completion of the human map

From the comparative mapping studies carried out thus far, it is clear that there are large chromosome segments which show conservation of linkage (synteny) and order of genetic markers between mice and humans (and other mammals). On average these conserved blocks appear to be of the order of 10cM but in some cases almost a whole chromosome may show conserved synteny. Hence, if a region of a mouse chromosome is mapped to high resolution, the information can be used to make predictions about the homologous region of the human genome and *vice versa*. Mapping information in particular mouse chromosome regions has helped to clear up ambiguities in the human map. The present relatively low resolution comparative map is likely to be misleading. Once more markers are placed on particular chromosome regions it may become apparent that there has been more scrambling of genomes than suspected from the low resolution map.

It is likely that regions that are refractory to cloning from humans, may be amenable to cloning in the mouse and *vice versa*. The comparative approach will thus facilitate the completion of contigs and chromosome walking exercises towards disease genes. Furthermore, comparative analysis of cosmids and YACs containing inserts from homologous regions of the mouse and human genomes may help in the identification of coding regions as well as regulatory elements of genes.

Using the European backcross (see 3.3.4) it will be possible to obtain a very high resolution genetic map (average spacing of 0.1cM - 200kb). Markers used to create this map can then be used to identify YACs which can be incorporated into a contig of the mouse genome. In addition, long range physical mapping using pulsed field gel electrophoresis can proceed more quickly and efficiently where high resolution genetic maps already exist. The genetic and physical mapping should proceed together. The mouse map should be able to keep pace with the human map provided sufficient resources are made available.

2. *Mapping mutant loci in mouse and the mapping of human disease loci*

a) Single gene disorders.

There is a large number of human single gene disorders which are very difficult to map to chromosomes or to precise chromosomal regions due to the unavailability of informative families or due to heterogeneity. Particular forms of deafness and retinal degeneration are examples of such disorders. In mouse, a number of deafness and retinal degeneration mutants have been described which have already been mapped or are in the process of being mapped. Given the improved resolution of the mouse map and the availability of microsatellite markers this is now becoming a relatively simple exercise. Once these loci have been mapped in the mouse it will be possible to use the knowledge of the comparative map to predict the chromosome regions in humans.

In mouse it is possible to map mutant loci to high resolution as any particular phenotype can be segregated in a very large (eg 1000 animals) backcross. This would allow markers to be placed within 200kb of the mutant gene, thus making gene identification a relatively simple task. This contrasts with the human situation where many mutations will be placed within regions of 10Mb or greater.

Many more mouse mutants are being created using homologous recombination, gene traps or insertional vectors. Ultimately these mouse models will be an enormous aid to unravelling the basis of the diseases and to testing new therapies.

It is important to add that a similar phenotype in mouse and man does not mean that the same genes are involved; even in these cases the mouse mutant, though not a direct model for a human condition, will still reveal insights into developmental and regulatory mechanisms.

b) Multifactorial diseases.

Many of the major diseases affecting humans may have a genetic basis but are intractable because of their polygenic nature, because of low penetrance and because variable environmental factors play a major role. Hence, major cancers, heart disease, hypertension, diabetes, obesity and epilepsy are likely to have a polygenic/multifactorial basis. Mouse (and rat) models are proving to be very powerful systems for studying the genetic basis of these conditions and for mapping the key loci involved. A number of animal models exist for these disorders, which can be either multifactorial or caused by a single gene defect (for example, there are single genes which give rise to both obesity and diabetes in the mouse). These single gene models present more tractable cloning problems, and may help identify genes which are important as a component of the multifactorial disorders. Once the individual predisposing loci have been mapped in the mouse the comparative map can be used to predict human regions involved and these can then be tested. This approach is already being applied for diabetes and hypertension.

3.3.3 *Mapping of cDNAs on the genetic linkage map using the European backcross*

There are now considerable efforts to isolate and sequence random cDNAs as a rapid route to identifying genes. Of the order of 30,000 different cDNAs have already been partially sequenced. It is essential that these be mapped. In humans, most of these cannot be placed on the genetic map because they will not detect polymorphism. However, nearly all the cDNAs (even human ones) can be mapped to the European backcross directly. Once cDNAs have been placed on the map their positions can be related to microsatellite markers which have in turn been used to map mouse mutants. This will immediately present candidate genes for mouse mutants, and, in some cases human diseases. This is a very important advantage of the mouse relative to the human. These cDNA clones will be mapped to human chromosomes and to regions of chromosomes by physical methods, so rapidly enhancing the comparative mapping exercise. If the UK takes a coordinating role in this exercise, the UK genome mapping community would have a unique advantage.

3.3.4 *Constructing a high resolution mouse map - the key role of the European backcross*

The UK has played a central role in initiating and coordinating the large European backcross. This is a thousand animal interspecific backcross between two strains of mice *Mus musculus* and *Mus spretus*. The critical feature here is that just about any type of DNA probe (ie cDNA, random single copy DNA fragments or microsatellites) will show variation between these two species and can therefore be mapped on the backcross. This backcross is much larger than any previously created (usually only 100-200 animals) and will therefore allow a map with much higher resolution.

So far at the MRC HGMP Resource Centre, three to four anchor markers (often, but not always, genes) have been mapped per chromosome in the backcross. The European backcross is now available to the community. Markers of interest to users can be mapped at the MRC HGMP Resource Centre.

This initial mapping exercise of anchor loci has been extremely successful but now it is time to crank up the process. It is crucial to map both microsatellite markers and genes to the backcross. Microsatellite markers are easy to use by PCR and most importantly can be transferred to other laboratories who wish to map mutant loci in inter or intraspecific backcrosses. However, microsatellites are not usually conserved between species and will not be useful for the comparative mapping exercise. Also, unless they are embedded within known genes, they have no biological information content. Thus, it will be vital to map as many genes as possible to the backcross. By relating the map positions of these to microsatellites it should be possible to pick candidate genes for particular mutations. The genes will of course be used to build up the comparative map.

Markers of several types will be applied to the backcross. These include:

(i) Microsatellites - in collaboration with the Whitehead Institute, USA - it is proposed that 3000 will be mapped to the backcross in the USA, 3000 in the UK.

(ii) Genes/cDNAs

(iii) CpG islands

(iv) Intra-repeat sequences - being isolated and developed at the ICRF. These should provide a cheap and efficient alternative to microsatellites but will not be appropriate for intra-specific crosses.

3.3.5 The physical map - interface with the genetic map

In principal, there will be sufficient recombination intervals in the European backcross to place an ordered array of markers every 200-300 kb on average. This would require in the order of 15-20,000 genetic markers. It is vital to use these markers that have been mapped to the backcross, to screen YAC libraries and other large insert libraries. This would go a long way to producing a complete contig and genetic and physical maps could be built up side by side. The concern here is that, although the UK and Europe are playing a strategic and controlling role in the backcross exercise and in constructing the genetic map, the physical map/contig assembly will be taken over by the group in the USA which is better resourced. The UK may miss out in the exploitation at the end of the project just as its application becomes important.

3.3.6 Is there a need for a mouse genome centre?

The UK has the personnel and expertise to make a major contribution to the mouse genome initiative and comparative genome analysis. There is some concern that full advantage is not being taken of the UK position. A crucial issue is whether there should be a mouse genome centre with sufficient resources to produce and coordinate the genetic map and to carry out the YAC contig/ physical mapping exercise. A centre of this sort could also use the information obtained to help map and clone mutant gene loci.

The second issue is the location of such a centre. It would be an advantage for such a centre to be located close or adjacent to a strong centre of human genome research.

There are already a few key individuals in the UK who are creating mouse genetic maps, positionally cloning genes and developing the resources and automated technology for constructing the physical map. Involvement of these individuals is likely to be crucial for the success of the enterprise. Clearly this issue has relevance for the future directions of the MRC Radiobiology Unit at Harwell, which is a major centre for the identification and characterisation of mouse mutants, and the MRC HGMP Resource Centre.

3.3.7 *Preserving and supplying mouse mutant stocks*

One important issue is the preservation and availability of mouse mutant stocks in one form or another (eg as frozen embryos). Several interesting mutants already seem to have been lost without trace. This is clearly becoming critical as hundreds of new mutants are being produced per year through new technologies. The Jackson laboratories in the USA have set up a repository of mutant stocks which is available to researchers at a reasonable price. The question is whether the UK/Europe should also set up such a bank of mutant stocks as they are being generated. Embryo banking facilities at the MRC Radiobiology Unit and the MRC Experimental Embryology and Teratology Unit are currently under consideration by the MRC.

3.3.8 *The rat*

Genetic analysis in rats is far behind that in mice because of the relative cost of breeding programmes and the development of transgenesis in mice. For many years rats have been the mammal of choice for physiological, neurological, pharmacological and biochemical analysis. It is now appreciated that rats may provide genetic model systems for human complex disorders with a physiological and neurological basis, for example, hypertension and epilepsy. For various reasons, there are no mouse models for these conditions. There are now programmes to map loci associated with predisposition to hypertension. Crosses have been set up and several hundred rat microsatellite markers have been developed.

There is very little coordination of the rat programme. Unlike the mouse, there are no comprehensive lists or databases of complex and monogenic traits in rats. Thus full advantage is not being taken of the possibilities offered by the rat. There is now a demand for markers but no system for distribution of these markers. This may be an issue of relevance for the MRC HGMP Resource Centre.

3.3.9 *Farm animals*

For obvious reasons the human and mouse genetic maps are by far the most advanced and most intensively studied of mammalian species. However, there are initiatives to map the genomes of domestic farm animals including sheep, cow, pig and chickens. The main impetus is to map and clone the genes associated with commercially important traits, for example, in the pig these include growth, backfat and fertility.

At present, the avowed and sensible aim of these projects is to construct relatively low resolution maps with markers placed at approximately 20cM intervals. This may be sufficient to map the traits to chromosome regions. The refined mapping and ultimate cloning of the commercial trait genes are likely to draw heavily on the much more advanced human and mouse maps. Thus the domestic animal programme will be a major beneficiary of the human genome programme.

Human genetics will also benefit from the genetic analysis of farm animals in two ways. Firstly, an understanding of gene linkage and order in different mammals (and birds) will provide important insights into genome evolution. Already it seems clear that there has been much less scrambling between the cattle and human genomes, for example, than between the mouse and human genomes. Secondly, traits being mapped in farm animals may also be relevant for human health and biology.

The farm animal mapping programmes are mainly international in character but for certain organisms individual countries dominate. For example, the USA and Australia are taking the lead in the cattle programme, New Zealand is prominent in the sheep programme. The UK is playing a minor role in these programmes but is prominent in both the pig and chicken mapping programmes. For example, the first chicken linkage map, involving 150 markers has been produced at Compton.

There are two issues worth raising in the context of this document. Firstly, the UK farm animal mapping programme appears to have been relatively under-resourced with most research staff on short-term contracts. This compares with the French programme which has 10 permanent staff employed on the porcine mapping project. However, the AFRC in its plant and genome analysis programme recently made awards of £2m, of which £0.8m was for work on farm animal genomes.

Secondly, backcross mapping in farm animals is a difficult and cumbersome process. Problems are being encountered in carrying out a large bovine backcross. In addition, the maps are being constructed with microsatellites which do not contribute to the comparative map. It would be extremely helpful to have an efficient and rapid physical mapping approach which can be applied to all farm animal genomes and to all types of markers including genes.

3.3.10 The pufferfish

The pufferfish, *Fugu rubripes*, is a vertebrate which has a genome approximately 7.5 times smaller than that of the human. The introns in pufferfish are very much smaller than those found within the human genes. Furthermore, there appear to be no interspersed highly repetitive sequences and as yet no pseudogenes have been documented. Hence genome analysis in pufferfish will be much easier than in humans; it should be much easier to identify genes - and possibly regulatory regions as well.

There is a further exciting possibility. If gene order is conserved at all in *Fugu* compared to that in humans this should facilitate the positional cloning of genes important for human disease. To be useful conserved linkage need only extend across a stretch of five or six genes.

Hence *Fugu* holds considerable promise as a model vertebrate for genome analysis. For the time being Dr Brenner's group in Cambridge is carrying out pilot studies which will allow an evaluation of *Fugu's* full potential. If the results of these pilot studies are encouraging an expansion of the project must be considered. In the short term, if necessary, resources should be made available to allow distribution of large insert genomic pufferfish libraries to the community.

3.3.11 Recommendations

(i) The establishment of a mouse genome centre is urgently required so that advantage can be taken of the strength of the UK in this area. The centre should include expertise in genetic and physical mapping as well as positional cloning. Ideally, the centre should be adjacent to an existing mapping centre to ensure good cross fertilisation of ideas for technology development.

(ii) Priority should be given to the coordination of mapping of cDNAs on the human and mouse maps.

(iii) Strong support for the worm *(Caenorhabditis elegans)*, the fruit fly *(Drosophila melanogaster)*, yeast *(Saccharomyces cerevisiae)* and plant *(Arabidopsis thaliana)* projects should be continued.

(iv) The rat is an important model for pharmaceutical research and is the organism of choice for studying many aspects of whole animal physiology, brain function and polygenic disease. A coordinated programme between the mouse and the rat should be investigated.

(v) The strategic importance of mapping farm animals is very high. Support should be continued for European-wide cooperation to build on existing programmes. The level of support within the UK programme should be investigated.

(vi) Integration, communication and technology transfer between different genome projects is essential. Mapping studies in other organisms with strategic potential such as pufferfish should be monitored and pilot studies initiated where appropriate.

3.4 DNA SEQUENCING

Of all the data that can be obtained from a genome, DNA sequence is the most fundamental. It is also the most durable: once a sequence has been determined to high accuracy, it remains of value into the indefinite future. The sequence is the ultimate map of the genome.

3.4.1 Genes or the whole genome?

In the case of the human genome, it is presumed that the value of the sequence varies substantially from region to region. The coding regions are obviously important, because they are translated directly into protein. The coding sequences of a given gene are broken up into many pieces (called exons) separated by largely unused introns (non-coding sequences of a gene), and at present are very difficult to recognise in their entirety by examination of genomic sequence. In front of, within and sometimes behind coding regions lie control sequences, which are at present mostly impossible to recognise without functional tests. Much of the remainder (probably the greater part of the sequence) is likely to be of less value in understanding the control of gene expression. Many researchers therefore believe that the best way of approaching the subset of DNA that is directly relevant to understanding gene expression within the human genome is to select entry points that are associated with genes. The following are some of the more important methods:

- cDNA sequencing. A cDNA is derived from the mRNA of a gene. It contains the coding sequence already spliced together into a single unit. mRNAs vary greatly in abundance in cells and a means must be found for sorting out a unique set of the cDNAs in advance of sequencing, if a large proportion of the genes are to be characterised in this way. Sorting can be achieved in part by using genomic entry points, such as the following.

- Exon trapping. Exons can be cloned from genomic DNA by taking advantage of their ability to splice onto other exons, provided by the experimenter, and the joined products can subsequently be isolated. Each such trapped exon provides an entry point to a gene. The abundance of exons is not related to their level of expression, so the sorting problem is less severe than for cDNAs.
- CpG islands. Many genes (about 60% is the estimate at present) have a CG rich region at their leading end. Unusually in the genome, the CpG dinucleotides in these regions are non-methylated. Means have been devised for isolating these CpG islands in large enough quantities to provide entry to all the genes that carry them.
- Tag sequencing. The idea is to acquire a short stretch of sequence from one end of a cloned cDNA or exon, in order to characterise the cloned segment. High accuracy is not required for matching purposes, so large numbers of clones can be handled and stored for future use. The method is being applied on a large scale to cDNAs.

Much of the information that is needed for biological understanding and for medical applications will be acquired in these ways. However, the data will always be fragmentary and incomplete. Investigation of a gene usually entails some local genomic sequencing, around the entry point, but extensive sequencing of intergenic regions is not required.

Full genomic sequencing, on the other hand, aims to collect all the information, both useful and useless. Thus nothing is missed, although at first most of the data will be uninterpretable without assistance from more biologically oriented approaches. By this approach, the overheads of sorting entry points, the difficulty of dealing with unusually large or variably spliced genes, and additional sequencing to cover control regions, are avoided. As noted above, the sequence forms a permanent archive of increasing and cumulative value as analytical methods improve. While it is desirable that all genomic information is placed in the public domain, this is particularly important for genomic sequence because so much of its value lies in the future.

The importance of the human genome is such that eventually it may be sequenced in its entirety - at first piecemeal, as required by individual investigations, but later as a concerted effort. It is too early for resources to be applied to the whole genome for this purpose, but megabase scale sequencing is proceeding in certain high value (gene-rich) areas of the human genome, as well as in a number of simpler (so-called model) organisms. In the latter, the genes are more tightly packed and more easily recognised; thus the breakeven point, where the falling cost of sequencing matches the value obtained, has come earlier for these systems.

It is likely that sequence analysis of the genes of model organisms will provide important information on the major types of protein used in living systems. It is already clear that the major differences between organisms result not so much from the types of gene each organism contains but more from the way the genes are used- the time or position a gene is first switched on in development, for example. Thus there is reason to believe that the majority of genes found in simpler organisms will be found in the human and that there will be relatively few truly novel classes of genes in humans. The identification of this set of 'fundamental' genes will be complete within a three to five year period and will be a very important contribution to human biology.

3.4.2 *Innovation anticipated in the next five years*

There is a good chance that primary large scale sequencing will continue to depend on the random termination/size separation method of Sanger. Detailed improvements in sample handling and data assembly, with all possible steps being automated, will deliver at least a two-fold decrease in costs. This process has already begun, with the introduction of effective fluorescent sequencing devices; these will become cheaper to run, and will be incorporated into fully automated systems. An interesting variant is the multiplex method, in which samples are mixed together for ease of processing and then sorted out by hybridisation; at present the automation of this procedure seems more difficult, but this situation may change.

The new method of sequencing by hybridisation (SBH) is likely to become very important in diagnostics, but its role in primary sequencing is less certain.

Sequencing by insertion is very new and less advanced: its status cannot yet be judged. Other novel methods, including mass spectrometry, single molecule degradation, and scanning tunnelling microscopy, are very distant contenders.

3.4.3 *Organisms sequenced*

A very wide range of organisms have contributed some sequence to the databases, at the level of individual genes, but in only a few are large scale programmes, in either cDNA or genomic sequencing, complete or underway worldwide.

- complete: viruses (Φx174, adenovirus, herpes virus)
- current: bacteria *(Escherichia coli)*, yeast *(Saccharomyces cerevisiae)*, worm *(Caenorhabditis elegans)*, fruit fly *(Drosophila melanogaster)*, human.
- starting: numerous microorganisms *(Bacillus subtilis, Salmonella typhimurium, Mycobacterium leprae, Mycobacterium tuberculosis, Mycobacterium capricolum), Dictyostelium discoides, Arabidopsis thaliana, Fugu rubripes,* mouse.

3.4.4 *Source of material for sequencing*

Genomic sequencing usually begins from a genome map - ie a collection of overlapping clones that have been arranged in order so as to completely cover the region of interest. At present, the most convenient clones for sequencing are cosmids, which hold some 40kb of DNA apiece, but other cloning vehicles, that hold larger fragments, are coming into use (P1, BAC). YACs, which can hold >1Mb, are exceedingly valuable for long range mapping, but are less convenient for sequencing because of the difficulty of assembling large pieces containing repetitive DNA. Usually YACs are converted into smaller pieces, either by subcloning or (preferably) by selection of clones from an independent library (see Section 3.2.3).

3.4.5 *Cost*

At present, the cost of complete sequencing is approximately $1 per base for genomes of moderate complexity (eg worm), but probably $2 for human. The principal cost (and the reason for the discrepancy) is that of 'finishing' - ie not the collection of raw data, but the process of sorting out problems and ensuring that the sequence has been fully determined on both strands. Repetitive sequences (both direct and inverted) and heavily biased base compositions are the main difficulties, resulting in either logical conundrums or enzymic failure. It is on efficient resolution of these problems, as much as on cheaper collection of raw data, that the way forward depends.

3.4.6 *Financial support*

In the past, the MRC has funded the sequencing of the yeast and worm genomes at the Laboratory of Molecular Biology in Cambridge. Currently, the MRC provides major support for large scale sequencing in worm and some sequencing of human chromosomes, and the Wellcome Trust provides major funding for yeast and human sequencing: these projects are united in the Sanger Centre. The worm *(Caenorhabditis elegans)* is something of a flagship project. The *Arabidopsis thaliana* sequencing project is supported by the AFRC, and a European sequencing network is being set up (as was done for yeast). Most other genome sequencing is carried out on a smaller scale as part of specific research programmes, and needs no separate justification.

UK initiatives in cDNA, exon and CpG island sequencing need support and strengthening, in order to ensure that we do not lose the more immediate gains to be had by these means. Developments in sequencing by hybridisation need national support if they are not to be lost to the USA.

3.4.7 *Recommendations*

(i) Priority should be given to fund the sequencing of gene-rich regions of the human genome.

(ii) All sequence data should be placed in the public domain.

(iii) Genomic sequencing projects for model organisms with well developed genetics should be supported. This sequence analysis will provide important insights into the biology of living organisms.

(iv) Priority should be given to cDNA, exon and CpG island sequencing in order to ensure that the UK does not lose the more immediate gains to be had by these means.

(v) Large scale DNA sequencing will continue to depend on the method of Sanger. However, it is important to invest in the development of new technologies. Sequencing by hybridisation will become important in diagnostics and its role in primary sequencing should be investigated.

3.5 GENOME INFORMATICS

The main product of direct genome research is information, either in the form of relative map positions, or ultimately as DNA sequence. This information is an invaluable resource for both basic and applied research. The key elements of genome informatics are therefore the collection of data, its analysis and organisation into structured databases, and its distribution to biological researchers. However, these activities do not just involve data pushing; there is a need to develop new algorithms and techniques to automate collecting and assembling the information, and also to help identify potentially important items in the wealth of finished data (eg gene candidates). So there is a significant research component to informatics, in addition to the provision of a service. A significant lesson that has been learnt is that it is not easy to apply existing computer software technology to genomic problems, for example to genomic mapping databases.

3.5.1 *Data collection*

Data come from both large and small scale operations. Increasingly, the large systematic data sets that will make the backbone maps and sequences are being gathered by large centres that make heavy use of automation and on-line data collection. These groups tend to each develop their own data collection software. Although wasteful, this happens because it is an integral part of their technological development environment, over which they require close control.

On the other hand, smaller laboratories cannot afford the resources for software development, and must either use very simple computer systems (PC spreadsheets are often satisfactory), or a system developed by a larger centre.

The alternative of a third party computing group developing software for data collection and management does not appear to have been successful. Almost all programmes that have been successfully adopted have been developed in the context of a large data gathering group. Note that this is not necessarily the case for more analytic programmes involved, for example, in sequence analysis, which tend to come from specialist research groups.

In summary, development of data collection and assembly software should be supported as an integral part of large scale data collection projects. This normally requires full-time input from a senior individual, perhaps with assistance. To make software transfer as easy as possible it is desirable that programmes should :

(a) be required to be easily available, with some degree of support for serious new users by the developers, and

(b) be as flexible as possible, since systems almost always need some modification for a new environment. This means that the source code and data structures should be available, which conflicts with traditional approaches towards software exploitation.

3.5.2 Data management

There are two levels of data management. One is private, within the laboratory, and the other public, providing a canonical database. The ideal is to be able to transfer information easily between them: new results from laboratory to public database, and relevant results from the public database into the local environment. This requires standards, both of data structures, and of quality and consistency of data.

The main public genome databases are at EMBL in Heidelberg and NCBI (National Center for Biotechnology Information) in Washington DC for sequences, and at GDB (Genome Database) in Baltimore for human genetic maps. At the moment the sequence databases are much more widely used than mapping databases.

The sequence databases are mature and in reasonably good shape. Following problems in the late 1980's, EMBL in Heidelberg, Germany, and Genbank in Los Alamos, USA, now both keep up to date with incoming data, and are both planning for the future. The EMBL database will come to the new EBI (European Bioinformatics Institute) in Cambridge in 1994, which will allow the EMBL database staff to provide better on-line access to data, as well as additional facilities as a resource centre, particularly for training and research. However, at least for the first few years, they will continue to concentrate on DNA and protein sequence, and not on genetic mapping data.

The human genome database, GDB, is currently based on loci, with poor map facilities and a lagging update record. As a clone based human physical map is assembled in the near future, it is going to be essential to have a well maintained and accessible physical map database. Without this, the data will be of marginal use until they are published in final form. Experience with model organisms has been that physical map information is useful long before this point, but that it must be available electronically, since only this allows continual updating and correction, and, ideally, can allow users to judge conflicting data for themselves. GDB is planning to provide physical map functions, but its success and future are uncertain.

The major problem with large databases is curation, ie ensuring that data is gathered systematically and stored in an orderly fashion. This is best handled by one dedicated individual, but all the major genomic data sets have grown too big for this. One current trend towards resolving this is to maintain a 'backbone' database, containing unambiguous data, such as DNA sequence, well established clone contigs, or well ordered sets of genetic markers. In conjunction there would be multiple distinct 'annotational' databases that refer to the backbone, which therefore links them together. Each component is easier to maintain than a combined whole, and a type of free market can apply for the annotational databases, whereby it is easy to start up a new one.

The major disadvantage of the backbone approach is the effort required to keep track of all the pieces. One solution is to have an overview structure that brings the pieces together by copying them from the people that curate them, providing a uniform view. There is a new EC project, IGD (Integrated Genome Database), which is based at DKFZ (Deutsche Krebsforschungs Zentrum) in Heidelberg and has strong UK involvement, that is aiming to do just this for all human map and sequence information. A comparable attempt in the USA is the ASN.1 system from NCBI, which is currently oriented towards sequence and bibliography.

The trend towards backbone databases should be supported; groups building local databases should be strongly encouraged to connect into them (eg via IGD). It is very important that the primary sequence and map data from large projects is properly integrated into the backbone. Without that it will be of little value.

For the backbone there is an urgent need for a reference database of record for physical and genetic maps, ie either a major revision of, or a replacement for, GDB. It should not try to do too much (eg it should not handle complete annotation). For more detailed information, various groups are developing chromosome specific databases, including X and 11 in the UK. These require a long-term commitment to curation, and also new software.

It has become clear that standard database software (which uses the relational model) is inadequate for representing the full richness of genomic data. Various alternative systems are under development, including ACEDB from the *Caenorhabditis elegans* project in Cambridge, an extension of the RLDB (Reference Library Database) programme at ICRF, and a number of systems in the USA. Software development is always time consuming and risky, but is necessary in this area. Note that the backbone database, which must have absolute and continuous integrity, but is of more limited scope, should use sound established software.

3.5.3 Distribution and hardware

First and perhaps most importantly, good wide area network connections are essential. Data must be made available either as a local copy, or across the national and international network. The advantage of having a local copy is that local networks are much faster, but there is a management problem in keeping such copies up to date. Although some large standard databases can be maintained by CD-ROM, a good network connection allows regular updating, and access to less frequently used databases. International networks are particularly important, since many of the basic databases will be maintained abroad, in the USA, France, Germany or Japan. The basic international network uses the Internet Protocol (IP), which is now available (as JIPS) over the UK academic network, JANET. Currently the international connections are much inferior to national ones. JANET connections are better than in most of Europe, but typically worse than in the USA. JANET is currently being upgraded to Super JANET and it is important that the community has good access to Super JANET.

As well as accessing data, network connections also allow access to types of machines and programmes not available locally. This is the principle adopted by the MRC HGMP Resource Centre, which provides accounts on a cluster of Unix workstations. They also run a successful series of courses in the use of the packages that they make available. Running programmes across the network, particularly graphical programmes, which are the most natural for representing map information, requires especially good network connections. To maximise their potential, it is important that both the MRC HGMP Resource Centre and the EBI have first class connections to the Super JANET core network. The MRC HGMP Resource Centre should continue to provide access to as many databases and packages as possible, while also aiding individual sites with the appropriate resources to install copies of the most frequently used software, so as to reduce load on its own computers. The provision of a central computing centre should not prevent the funding of local workstation level machines for groups with a sufficient workload, since they ultimately provide greater flexibility.

The MRC HGMP Resource Centre's Informatics Group currently has 1700 regular users of its service out of 2000 registered users with the Centre which, based on present costs, equates to approximately £550 per user, per year. On completion of the equipment replacement programme currently being considered by the MRC at a cost of approximately £200,000, it will be possible to accommodate additional users at a concomitant reduction per user in annual costs. In order to speed up the network links to the MRC HGMP Resource Centre it will be extremely important to link into the Super JANET System as soon as practicably possible - the cost of such a connection and the first year's service is estimated to be in the order of £200,000 (to be shared between the MRC HGMP Resource Centre, Sanger Centre and EBI at Hinxton Hall).

3.5.4. *Recommendations*

(i) All data should be deposited in centrally funded databases. Access to the databases should be through electronic networks available to both academic and industrial users. The MRC HGMP Resource Centre and the European EBI must have very high speed network connections.

(ii) A high quality of database curation is important to make maximal use of data. Support should be made available to UK groups curating subsets of the genomic data. Database development is required and should be supported in the context of these groups.

(iii) Communication between the databases used for different organisms should be coordinated.

(iv) The state of the international central data repositories is of concern. The UK should maintain active involvement with planning for both the sequence libraries (the European copies of which are managed by EBI) and the mapping database, GDB. A strong European component of/counterpart to GDB should be encouraged.

(v) Software development is critical for large scale data collection, and this should be supported in conjunction with large scale sequencing and mapping projects. This software should also be usable by smaller groups.

4. COMMERCIAL OPPORTUNITIES

4.1 INTRODUCTION

The commercial implications of the information being derived from the HGMP are considerable. The implications are not all long term. They are analogous to those arising from the discovery of recombinant DNA technology. There will be both direct and indirect commercial opportunities for the biotechnology and pharmaceutical industries.

Most drugs which are used to treat common diseases treat the symptoms of the disease rather than altering the underlying pathology. The most obvious examples of this are the drugs used to treat rheumatoid arthritis, beta adrenoceptor agonists used to treat asthma, and cancer chemotherapy and radiotherapy which both affect rapidly dividing cells - a feature of some tumours. An exception to this general rule is infectious disease (whether of bacterial or viral origin) where the drugs developed (eg antibiotics) are specifically targeted to inhibit the replication of the organism responsible for the disease.

The underlying molecular pathology of complex and common diseases is not well understood despite much biological and biochemical work investigating the phenotypic characteristics of the cell types involved both *in vitro* and *in vivo*. The 'positional cloning' technology developed to map monogenic disorders, using microsatellite markers, is equally applicable to polygenic diseases provided the right families and pedigrees or affected individuals are available. Like monogenic disorders, such as cystic fibrosis and fragile X syndrome, it will be possible to map genetically and then locate physically those genes which predispose people to certain diseases. Already it is known, for example, that overexpression of the gene for angiotensin converting enzyme (ACE) is a predisposing factor for myocardial infarction and that defects in the low density lipoprotein receptor predispose to atherosclerosis owing to high blood cholesterol levels. A gene which is responsible for resistance of certain mouse strains to *mycobacterium* infection has also been identified recently by microsatellite mapping.

The identification of predisposing genes is of fundamental importance because it is not only a means of understanding the pathology of the disease (or resistance to infection), but it will also provide novel targets for drug intervention or validate existing targets. If ACE inhibitors had not been developed, the uncovering of ACE as a genetic factor in heart disease would immediately have suggested this enzyme as a target for drug inhibition. So few genes and gene products are currently known that it is likely that completely new gene families will be discovered by positional cloning, providing a multitude of new targets for drug intervention in many diseases.

Table I
Common diseases with a recognised genetic component
1. Heart disease
2. Hypertension
3. Schizophrenia
4. Manic depression
5. Diabetes
6. Rheumatoid Artheritis
7. Cancer
8. Asthma/atopy
9. Obesity

Table II
Therapeutic implications of human genome project
1. Diagnosis of genetic diseases and predisposition to disease
2. Prevention based on diagnosis
3. Gene therapy
4. Protein therapeutics
5. Identification of novel points of intervention for small molecular discovery programmes
6. Robotics for high throughput screening

Table I shows some of those common diseases with a recognised genetic component and Table II some of the short and medium term opportunities that will arise from the discovery of predisposing genes. Again, it should be noted from Table II that the opportunities are not all long term. The uncovering of a genetic marker for a disease will allow new approaches to diagnosis: the importance of the diagnosis of genetic disorders and their management from a behavioural and social science point of view should be stressed. Genetic screening will also allow further clinical differentiation of diseases which are not homogeneous (eg type II diabetes). There is a need to consider the situation from molecular genetics to family practitioner. Education about and communication of genetics are both areas that need active support; much of this could come through government coordination.

The development of gene therapy and application of biotechnology to generate protein therapeutics will happen more quickly than anticipated. The genome project will provide many genes or gene products which may be useful therapeutically if suitable delivery systems are developed. A paradigm for this is cystic fibrosis. In the four years since the discovery of the gene defect responsible for the disease, gene therapy based on the cystic fibrosis transmembrane conductance regulator (CFTR) protein is being carried out. In addition, pharmacological approaches to treating the disease, based on the restoration of ion channel function or reversal of the conformational defect of the mutant protein, are being tried. Furthermore, the diagnosis of the three or four major CFTR mutations in Caucasian populations is now routine. Much of this work is being driven by small biotechnology companies, from a knowledge of gene structure and function.

4.2 EXPLOITATION

4.2.1 Background

The potential commercial importance of the HGMP is undisputed. The UK is particularly well placed to exploit the technology scientifically owing to a good molecular genetics base and a well developed clinical genetics establishment, with well documented families and access to pedigrees. The UK, however, is not well placed to exploit the technology commercially. With certain exceptions the UK has been slow to pick up on commercial opportunities in biotechnology and there is still much technology languishing in academic laboratories that has not been adequately exploited. This is largely an infrastructural problem. The USA model of technology transfer (and wealth creation) is based on the formation of small 'start up' companies around academic laboratories (or several related investigators) generally funded by venture capital. Usually the companies develop by attracting corporate investors into strategic alliances, followed by a private placement of shares with the companies, eventually obtaining wider investment and more capital by a public offering. This route is possible in the USA for several reasons - notably a well established risk orientated venture capital community, a wealth of opportunity in an educated and specialised investor community and a public interested in science and technology.

In this country we have a small and slightly risk averse venture capital community in a financial climate which is not sympathetic to small companies with reasonably long time frames (five to ten years) before products and profitability. With the new opportunities arising from the human genome project, this has to change. Technology transfer units in universities and the funding agencies are quite used to transferring compounds or reagents but generally have neither the experience nor the inclination to transfer technology by starting up new companies.

There are, of course, exceptions and some successes. The recently established Cantab Pharmaceuticals and Therexsys are both examples where the USA model has been followed. British Biotechnology, Celltech and Xenova started some years ago, are all considered to be, to a greater or lesser extent, successful biotechnology companies. They have certainly provided employment opportunities for highly qualified and able people with an entrepreneurial attitude. They are also likely to create wealth in the medium term. However, there is reluctance from the large UK based pharmaceutical companies to seed small start ups (from inside) and there is still considerable antipathy and misunderstanding of industry and commercial reality by the academic community of the UK.

In the USA there are already several companies formed to find and exploit genes which predispose to diseases and there is clearly considerable interest from some of the major pharmaceutical companies in interacting with such small companies in the genome area. SmithKline Beecham have recently announced an alliance with Human Genome Sciences in Bethesda and Glaxo has announced that it has started a human genetic initiative specifically to address the molecular pathology of common diseases.

For other companies, particularly those with an inadequate pipeline of new chemical entities, the genome project provides a real opportunity for innovation and for the discovery of disease modifying agents working at novel points of intervention.

4.2.2 *Exploitation mechanisms*

There are several potential ways of exploiting the technology. One of the most important considerations here is the potential intellectual property situation.

Patent protection in biotechnology in general and the HGMP in particular is a complicated issue. At least for the patenting of genes encoding therapeutic proteins there is now an established case law both in Europe and in the USA. Whether short DNA sequences obtained at random (EST's - expressed sequence tags), which specifically represent genes but which may not of themselves be useful (except perhaps as diagnostic tools), will be patentable remains to be seen. What is important however, if short sequences are not patentable, is that any useful gene which is subsequently discovered by some other route, is not rendered unprotectable by virtue of prior disclosure of part of the sequence from simple random sequencing.

The likelihood is that genes derived by positional cloning which meet patentable criteria (novelty, non-obviousness, inventive step, reduced to practice, etc) will probably be protectable. Any exploitation strategy must take this into account and draw the right line between competitive and precompetitive information. An international perspective is also required and the UK must not be parochial in its view.

Table III
Potential exploitation routes
1. Genome company(s) 2. Genetics consortium 3. Independent academic-industrial link 4. Entirely *ad hoc* 5. Multipartner research programme

Table III shows a number of possible mechanisms for exploitation. As described above the genome company concept prescribes the USA approach to technology transfer in the biotechnology sector. There are several advantages - not the least of which is the focus, enthusiasm and team work derived from people with a vested interest in the success of the endeavour. This is often a missing element in technology transfer. There are very clearly major internationally competitive academic groups around whom such a company could be crystallised.

An alternative to a new genetics company would be to form a consortium of the main academic groups in this area, plus the funding agencies and interested companies. As an example, a pharmaceutical company interested in understanding the molecular pathology of diabetes would become part of the consortium and would contract it to discover the genetic loci (by microsatellite mapping), to map the gene(s) involved in the disease (by positional cloning and sequencing) and to deliver the information to the sponsoring company. Provided means could be found to maintain confidentiality, this might be a cost effective and efficient way for the UK and USA companies to find 'disease' genes. It is anticipated that, much like a small company, the consortium would have its own laboratories and management structure where most of the work would take place. An analogy would be the MRC Collaborative Centre, although in this case there would have to be buy-in from all the key groups because they represent different parts of what is a complex technology (ie genetic mapping, physical mapping and high throughput sequencing). One considerable disadvantage of the consortium approach is that there is a tendency of these types of arrangements to become bureaucratic and ponderous. It will also require a good deal of flexibility and political good will to make it happen.

Several other mechanisms are also possible. In all of these it will be important to leverage from the academic groups all the different aspects of gene hunting - including genetic and physical mapping (and sequencing) and bioinformatics.

Multipartner research programmes are an approach but the intellectual property situation is likely to be more complicated than with a consortium or a company and confidentiality will be more difficult to keep. Clearly *ad hoc* or independent tracks are possible and will certainly happen if no coordinated approach is put in place quickly; indeed they are already starting to happen.

Government could play an important role in helping the interactions between academia and industry by providing seed capital for start-up companies or helping to protect the venture capital community and to promote their involvement. There is also the more general level where Government could act as the corporate midwife at the academic/industry interface. It should be possible to do this without compromising political considerations. Cooperative LINK-like schemes focused on cutting edge high quality genome mapping projects should be further promoted. Such schemes should be implemented in a manner which involves the best groups in a way that reinforces rather than detracts from their contributions to basic knowledge. A model scheme might be the MRC Protein Engineering LINK initiative at Cambridge.

Some companies already have in-house genome mapping programmes while others rely entirely on collaboration with the academic community. However, it is not always easy for industry to gain access to up-to-date genome information. Often, academic laboratories are working at the leading edge of genome mapping and do not wish or have the time to exploit the technology and/or data generated. A networking mechanism to improve the interface between academic and commercial laboratories is needed.

4.3 Recommendations

(i) The joint exploitation of genome research by funding agencies should be facilitated. The development of technology transfer from the funding agencies to the public domain should be encouraged. A clear statement from each agency on their strategy for the commercial exploitation of the research they fund is important.

(ii) There is an urgent need to investigate the ways in which the setting up of venture capital funded companies for gene discovery can be facilitated. Such companies should be able to access and exploit genetic and physical mapping technology in the UK or exploit it by another mechanism that leverages all appropriate technology. One possible way is to provide financial incentives for the setting up of small biotechnology companies.

(iii) A LINK-like scheme should be set up which builds on leading edge genome research. This will need to be implemented so as to involve the best groups in a way that reinforces rather than detracts from their contributions to basic knowledge.

5. EDUCATION AND TRAINING

5.1 Medical opportunities and the Health of the Nation

The most immediate practical impact of progress in gene mapping is in clinical genetics with new-found knowledge of disease genes and the mutations in these genes being applied to the diagnosis and counselling of inherited diseases. The identification of the causative mutations for monogenic disorders has already led to the availability of precise diagnostic tests, replacing less satisfactory statistical prognoses for families at high risk of these conditions. Virtually all of the more common inherited diseases are now amenable to genetic diagnosis, either by direct mutation analysis, or (in a few remaining instances) by linkage analysis. The ability to deliver these tests depends upon adequate academic and medical networks. The academic networks transfer research information into the clinic, sometimes in this field with dramatic rapidity, so that new techniques and gene markers are being used to give practical information to at-risk families within months or even weeks of their discovery. Medical networks are necessary to provide the proper patient care and counselling which underpin the application of these technologies. The Regional Genetic Centres in the UK have done very well so far, in taking up, developing and applying these techniques in a proper context, with adequate pre- and post-test counselling. There is also a research return on this, in that, families known to clinical geneticists through referrals for counselling have formed the research materials on which many gene-mapping studies were based.

So far, this activity has mainly been local with a strong R&D interaction. As more generalised tests become available, or as the disorders for which testing is required move from the relatively uncommon disorders into testing for predisposition to more common multifactorial disorders, the volume of activity will increase to the point that different models of service delivery will need to be explored. It is possible that this could be run from local pathology laboratories, or from differently constituted or expanded Regional Centres. It is also likely that this activity will become more attractive for private sector participation. Models that will allow both NHS and commercial development of large-scale testing, with adequate quality control, and adequate clinical interaction so as not to divorce the handling of laboratory tests from the very important delivery of pre- and post-test counselling, need to be urgently explored.

Experience gained on relatively uncommon monogenic disorders, both in testing high-risk families and in screening populations, will be invaluable models as knowledge advances on the genetic factors concerned with more common multifactorial disorders eg colon cancer, breast cancer, hypertension, ischaemic heart disease and the common psychoses. It is highly likely that screening will become both possible and desirable for some of these, as a means of targeting more expensive diagnostic technologies, or of instituting early clinical management. Many of the areas of most interest to clinical geneticists today, because of their significant genetic components, are also major targets in 'Health of the Nation', which lists among its key areas coronary heart disease and stroke, cancers and mental illness.

Because of the very rapid growth of knowledge in genetics over the past decade, it is likely to be necessary to introduce medical techniques into a situation where neither the professions nor the public are well educated in the underpinning science, its limitations, and its advantages. This leads to both unrealistic optimism, and to equally unrealistic fear of the new technologies. Urgent attention to public information in the realities of what can and cannot be expected of human genetics, at all levels from primary school to postgraduate medicine, and to the wider public, should be important components of planning to maximise the health impact of future discoveries from genome analysis. In the light of the White Paper, the MRC is reviewing its role in enhancing public awareness of science across its entire portfolio.

5.2 Training

There is a need to review the manpower requirements of the HGMP. A career development structure is particularly important because of the interdisciplinary nature of the project including, for example, biology, medicine, mathematical genetics and engineering. Special studentships and fellowships are already available through the GAHH initiative at the MRC but more action will be needed to ensure a reasonable supply of highly trained people in five to ten years time.

5.3 Recommendations

(i) Models which allow the development of large-scale genetic testing need to be explored urgently.

(ii) Education programmes in genetics should be set up at all levels from primary school through to postgraduate medicine in order to maximise the health impact of the HGMP.

(iii) There is a requirement for a careful assessment of the needs of industry, government departments and academia for individuals with training in genome analysis and its applications.

GLOSSARY

Base pair (bp): Two nucleotides, one from each of the strands of the DNA double helix, which are held together by weak bonds. Adenosine (A) pairs with thymidine (T) and guanosine (G) pairs with cytidine (C). 1 kilo base (kb) = 1000 bp, 1 Mega base (Mb) = 1000,000 bp.

BAC: Bacterial Artificial Chromosome, vector used to clone large fragments of foreign DNA.

Centimorgan: A unit of measure of recombination frequency; one centimorgan is equal to a 1% chance that a genetic locus will be separated from a marker due to recombination in a single generation. In man, a centimorgan equals on average, one million base pairs.

cDNA (complementary DNA): A DNA copy of the messenger RNA which corresponds to the gene expressed in the cell.

Chromosome: A rod-like structure composed of protein and DNA in the cell nucleus containing genetic information in the form of DNA; each human cell has 46 chromosomes in pairs. One chromosome of each pair is inherited from each parent.

Chromosome walking: The procedure used to move from one position on a chromosome to another. The walk is started with a DNA clone whose precise location is known and proceeds in a defined direction by generating successive rounds of overlapping clones.

Contigs: Groups of overlapping clones (usually cosmids or YACs) representing a continuous region of DNA.

Cosmids: Plasmids that contain specific sequences from the bacterial phage lambda (virus); cosmids are designed for cloning large fragments (typically 40,000 base pairs) of DNA.

CpG islands: A CG nucleotide rich region in the leading region of about 60% of genes. The CpG dinucleotides in these regions are non-methylated. Means have been devised for isolating these CpG islands.

Database: Repository of information.

DNA: DNA is normally a double stranded molecule, the two strands being held together by bonds between base pairs of nucleotides; it encodes genetic information. There are four such nucleotides, adenosine, guanosine, cytidine, and thymidine; base pairs only form between adenosine and thymidine and between guanosine and cytidine. It is therefore possible to determine the sequence of either strand from that of its partner.

DNA cloning: A means of isolating individual DNA sequences from a mixture of DNA and multiplying each to produce sufficient material for detailed analysis.

DNA sequence: The relative order of base pairs in a stretch of DNA, a gene, a chromosome or an entire genome.

EBV transformation: Immortalisation of a cell line using Epstein Barr Virus to ensure unlimited supply of the genetic material.

Expressed sequence tags (EST): DNA sequences that define gene products. ESTs are STSs derived from cDNAs.

Exons: The protein-coding DNA sequences of a gene.

Filter arrays: Cloned DNA immobilised in an ordered fashion on nylon membranes (filters).

Gene: The fundamental physical and functional unit of inheritance; an ordered sequence of nucleotides on a chromosome that specify the manufacture of a specific functional product such as a protein.

Genetic linkage map: A map showing the relative positions of gene loci on a chromosome; the distance is measured in centimorgans.

Genetics: The study of inheritance of specific traits.

Gene Mapping: Determination of the relative positions of genes on a chromosome and of the distance between them in linkage units (genetic mapping) or physical units (physical mapping).

Genome: All the genetic material in all the chromosomes of a particular organism; size is usually denoted in base pairs.

Genomic library: A collection of clones made up of a set of DNA fragments representing the entire genome.

Hybridisation: The joining of two complementary strands of DNA, or DNA and RNA, to form a double stranded molecule.

Informatics: The application of computer and statistical techniques to the management of information.

Interspecific backcross: Progeny generated by the mating of an offspring of two different species capable of interbreeding with either one of the parental strain.

Intraspecific backcross: Progeny generated by the mating of an offspring of two distinct strains of the same species with either one of the parental strain.

Introns: The DNA sequences that interrupt the protein-coding sequences of a gene.

Library: A random collection of clones whose relationship may be established by physical mapping.

Linkage: The proximity of two of more markers on a chromosome; the closer together the markers are the lower the probability that they will be separated during recombination. This gives an idea of the probability that they will be inherited together.

Marker: An identifiable physical location on a chromosome whose inheritance can be monitored.

Microsatellite: A stretch of di-, tri-, or tetra- nucleotide repeats flanked by unique sequences. The length of the repeat varies between individuals. Microsatellites are very useful for genetic mapping.

Multifactorial Disease: Diseases caused by variations in one or more genes and environmental factors.

Nucleotides: The building block of DNA or RNA; thousands of nucleotides are linked to form a DNA or RNA molecule.

P1: A bacterial virus used as a cloning vector for large stretches of DNA (typically 100 kb).

Phenotype: Physical manifestation of a gene product (eg eye colour).

Physical map: A map showing identifiable landmarks (eg genes, markers and RFLPs) on a stretch of DNA. Distance is measured in base pairs (bp), kilo base pairs (kb) or mega bases (Mb).

Plasmids: Circular DNA molecules found in bacteria that can carry foreign DNA of up to 12,000 base pairs. These have been extensively used in cloning genetic material in an readily accessible form, example for sequencing.

Polygenic disorders: Genetic disorders resulting from the combined action of more than one gene. The hereditary patterns of these disorders are more complex than single gene disorders.

Polymerase Chain Reaction (PCR): A technique that allows a sequence of interest in the genome (usually of the size of 50 bp to 2000 bp) to be amplified selectively against a background of the whole genomic DNA (in human 6 X 10^9 bp).

Polymorphism: Genetic variations in DNA sequence among individuals. Genetic variations occurring in more than one percent of a given population can be used for linkage analysis.

Positional Cloning: Isolation of genes by virtue of their position on a chromosome rather than from prior knowledge of the protein product.

Primer: Short artificially synthesized molecule of DNA to which new DNA can be attached using PCR.

Probe: Single stranded DNA or RNA of specific base sequence used to detect complementary base sequences by hybridisation.

Pulsed field Gel Electrophoresis: A method used to separate large DNA fragments (20 kb to 10 Mb) by applying pulses of current to the sample at various angles.

Restriction fragment length polymorphism (RFLP): Polymorphic variants in the size of the fragments produced by digesting DNA with an enzyme that cuts at specific sites.

Ribonucleic Acid (RNA): A molecule mainly involved in protein synthesis in a cell. It transforms the information contained in the DNA into protein.

Sequence Tagged Site (STS): A short DNA sequence readily located and amplified by PCR techniques, that uniquely identifies a physical genomic location.

Sequencing: Ordering of nucleotides in a stretch of DNA, a gene, a chromosome, or an entire genome.

Single gene disorders: Hereditary disorders caused by a defect in a single gene.

Trait: Any genetic character (eg hair colour).

Vector: A DNA molecule capable of autonomous replication in a cell and which contains restriction enzyme cleavage sites for the insertion of foreign DNA (eg plasmids, cosmids, P1, BACs and YACs).

Yeast artificial chromosome (YAC): Plasmids used to clone foreign DNA fragment inserts up to one million base pairs in yeast. Large fragments of DNA sequences can be maintained in YACs (250kb - 1 Mb).

ANNEX 1

Advisory Committee

Professor Thomas Blundell FRS,
Director-General, Agricultural and Food Research Council

Sir Walter F Bodmer FRS,
Director-General, Imperial Cancer Research Fund

Professor Sydney Brenner FRS,
Department of Medicine, University of Cambridge

Dr Peter Doyle CBE,
Director of Research and Technology, Zeneca Pharmaceuticals plc

Professor Malcolm A Ferguson-Smith FRS,
Department of Pathology, University of Cambridge

Dr Trevor Jones FPS FRSC,
Director of Research, The Wellcome Foundation Ltd

Dr Bridget Ogilvie ScD FlBiol Hon FRCPath,
Director, The Wellcome Trust

Professor Michael Peckham FRCP FRCPGlas FRCR,
Director of Research and Development, Department of Health

Dr George Poste,
Research and Development Chairman, SmithKline Beecham Pharmaceuticals

Sir Dai Rees FRS,
Secretary, Medical Research Council

Dr Geoffrey Robinson,
Chief Scientific Adviser on Science and Technology,
Department of Trade and Industry

Professor William D P Stewart FRS, (Chairman)
Chief Scientific Adviser, Office of Science and Technology

Sir Richard Sykes,
Deputy Chairman and Chief Executive, Glaxo Holdings plc

ANNEX 2

University Departments/Research Institutions

Professor Michael Ashburner FRS, Department of Genetics, University of Cambridge
Dr David Bentley, Head, Human Genetics Division, Sanger Centre, Cambridge
Professor Adrian Bird FRS, Institute of Cell and Molecular Biology, Edinburgh
Dr Steve Brown, Department of Biochemistry and Molecular Genetics,
St Mary's Hospital Medical School, London
Dr Martin Farrall, MRC Molecular Medicine Group, Hammersmith Hospital, London
Professor Sir Aaron Klug Hon FRCP FRS, Director,MRC Laboratory of Molecular Biology, Cambridge
Dr Hans Lehrach, Head, Genome Analysis Laboratory, Imperial Cancer Research Fund, London
Professor Alex Markham, Molecular Medicine Unit, St James' University Hospital, Leeds
Dr Anthony Monaco, Head, Human Genetics Laboratory, Imperial Cancer Research Fund, Oxford
Dr Michael Morgan, Programme Director, Wellcome Trust
Dr David Porteous, MRC Human Genetics Unit, Edinburgh
Professor Ellen Solomon, Department of Human Genetics, University College London;
Imperial Cancer Research Fund, London
Professor Edwin Southern FRS, Department of Biochemistry, Oxford
Dr Nigel Spurr, Head, Human Genetic Resources, Imperial Cancer Research Fund, Clare Hall
Dr John Todd, Nuffield Department of Surgery, John Radcliffe Hospital, Oxford
Dr Veronica VanHeyningen, MRC Human Genetics Unit, Edinburgh
Professor Sir David Weatherall FRCP FRCPath FRS, Regius Professor of Medicine, Institute of
Molecular Medicine, Oxford

Charities

Arthritis & Rheumatism Council for Research
British Diabetic Association
Imperial Cancer Research Fund
Motor Neurone Disease Association
Multiple Sclerosis Society of GB & NI

Advisory Committee (Annex 1)

Industry

Dr R Anand, Senior Scientist, Zeneca Pharmaceuticals plc
Dr D Bailey, Head of Molecular Sciences, Pfizer Research
Dr R G G Booth, Director of Biology, Roche Products plc

ANNEX 3

Questionnaire on Human Genome Research

1. Do you perceive any major gaps or overlaps in human genome research that need to be rectified ?

2. What applications, if any, does your organisation consider might benefit from the Human Genome Research Project ?

3. Do you think that there is adequate funding for Genome Research ? At present the Government support is directed through the Medical Research Council (MRC). In addition, funds are made available through the European Commission. It should be borne in mind that, in addition to Government support, some charities provide significant funding for the Human Genome Project eg the Wellcome Trust has funded a major initiative in sequencing and the ICRF has major initiatives in Physical Mapping.

4. Are the funding mechanisms for human genome mapping appropriate ?

5. Your views on Government's present strategy of funding human genome research would be welcome. Are there key areas that need urgent funding to facilitate the research, or are there other funding initiatives that need to be considered ?

6. The European Community has a human genome analysis programme as part of the Biomedical and Health Research Programme (BIOMED 1) which establishes a common base for the national programmes of community countries. A substantial amount of funding for research in the UK presently comes from this programme. Are there changes that you would wish to recommend/advise that the Government should be pressing for in future EC-funded programme arrangements ?

7. In your opinion what are the most appropriate strategies for the future of the UK Human Genome Mapping Project (genetic map, physical map, gene discovery) ? What research do you want available from the project ?

8. What initiatives should the Government be taking to facilitate interactions between academic and industrial human genome research?

9. Should the Government be playing a stronger role in the International field ie patenting issues ?

10. Are there areas of training that you consider need to be strengthened ?

ANNEX 4

1. ***Human Genome Research: A Review of European and International Contributions.*** Commissioned by the Medical Research Council, UK. by D J McLaren, January 1991.

2. ***Report on Genome Research.*** Commissioned by the European Science Foundation, 1991.

3. ***Research on the Human Genome in Europe and its relationship to activities elsewhere in the World.*** Academia Europaea, March 1991.

4. ***The U.S. Human Genome Project: The First Five Years.*** National Institutes of Health and Department of Energy, USA, April 1990.

5. ***A New Five-Year Plan for the U.S. Human Genome Project.*** F Collins and D Galas Science. Vol 262, p 43-46, October 1993.

6 ***Future Directions of Human Genome Project Considered.*** Human Genome News, NIH and DOE Vol 5, No 2, p 5-6, July 1993.

7. ***Realising our Potential - A Strategy for Science, Engineering and Technology.*** HMSO, Cm 2250, London, 1993.

8. ***Canadian Genome Analysis and Technology Programme.*** Genexpress, Vol 1, No 1, p 2-3, October 1993.

9. ***Genetic Screening - Ethical Issues.*** Nuffield Council on Bioethics, December 1993.